P9-CBM-653

WELCOME TO YOUR WORLD

SARAH WILLIAMS GOLDHAGEN

Welcome

to Your World

HOW THE BUILT ENVIRONME

SHAPES OUR LIV

An Imprint of HarperCollinsP

HarperCollins books may be purchased for educational, business, or sales promotional use. For information, please email the Special Markets Department at SPsales@harpercollins.com.

An extension of this copyright page appears on pages 299–307.

FIRST EDITION

Designed by Fritz Metsch

Library of Congress Cataloging-in-Publication Data has been applied for.

ISBN 978-0-06-195780-2

17 18 19 20 21   LSC   10 9 8 7 6 5 4 3 2 1

I dwell in Possibility—
A Fairer House than Prose—
More numerous of Windows—
Superior—for Doors—

—EMILY DICKINSON

Contents

Preface

This book comes with a bold promise. I, a stranger, welcome *you* to the world you live in every day. Yet I am confident that as you read what follows, what you know and how you think about your world will shift. It will become a different place than it was before you opened to this page. You will understand your role in it anew. You will realize that it affects you, your children, everybody in profound ways that you never imagined.

How do I know? Because it happened to me.

As a teenager, in the world predating smartphones and GPSs, I was fortunate to travel with my parents in Italy. During one long, searing, stressful car ride, as my father edged our rental car off the highway in the outskirts of Florence, my mother's sense of direction failed her, and we found ourselves lost. We had a map, but north could have been south or southeast or west, for all she knew, and nowhere on the map could she locate the names of the streets on which we were traveling. Stressed, my parents began to quarrel. Then suddenly, they and the car roiled and I found myself, not for the first time, caught in the middle of a titanic rage. Turn left! No, go straight! Look here, no, there, read that sign, that's not what it says . . . and on and on.

I insisted that my mother hand the map over to me. Quickly I identified our coordinates and told both of them, in a no doubt unpleasantly teenage voice, to be quiet. Solemnly, and pretty much flawlessly, I navigated our way to the hotel, where we registered our names in silence. Then I—desperate to get away, desperate to be alone—promptly announced I was going for a walk.

It didn't matter where; I just started walking. If my high school humanities curriculum contained lessons on Florence and its celebrated role in the Italian Renaissance, nothing of that surfaced. Surrounding me was just another Italian city, charming, but neither more nor less charming than many of the others we'd already visited. The path I followed seemed random, a product only of my wanderingly simmering distress.

The crowded street eventually opened into a plaza thick with honking cars, careering around buildings and tourists. A marvelous octagonal edifice was before me, sunken below the level of the street as if it was in the process of slowly burrowing its way back into the earth while also rising high above me. Marking out its height clearly was a grid of greenish-gray *pietra serena* stripes woven through the blocks of white marble on its facade. Behind this octagon—a Baptistery, I learned—an enormous cathedral extolled the glory of a defunct God while paying ecstatic tribute to the ingenuity of men. Nor was this all! Next to the cathedral's right flank, stretching skyward, was a lacy confection of delicate pinks, whites, and greens: Giotto's bell tower.

My heart soared. Serenity washed through my body. Within minutes, my liberation from the afternoon's torrid angst was complete. How could such beauty exist? Who created it? Why greens and pinks? How was it that three buildings in the sudden spectacle of a strange urban square could so utterly transform my mood, my day, and—little did I know it at the time—my life?

For much of the nearly forty years following that day, I've been writing about buildings, landscapes, and cities, first as a journalist, then as a professor and historian of modernism and its practitioners, including Louis I. Kahn, the American architect about whom I wrote a book. Ten years of teaching at Harvard University's Graduate School of Design immersed me in the architecture of today. My fascination with contemporary practices and ideas precipitated, in turn, an increasing frustration with academic publications' limited audiences and constrained forms, so I started contributing essays and reviews to general interest publications. For eight years I wrote as the architecture critic of the *New Republic* and by now have contributed to a bevy of scholarly and general publications here and abroad.

All this is simply to say that a good part of my professional and personal life has been devoted to trying to answer questions that I asked myself first on that memorable day. It's a journey that has involved traveling widely to explore and photograph buildings, landscapes, and cities, and reading deeply, as I investigated different ways of analyzing and thinking about the built environment. As an undergraduate and then a doctoral student in art history, I learned to appreciate the lasting power of visual languages and artistic traditions, and to consider how such traditions interact with both an individual designer's sensibility and a society's impetus toward innovative cultural expression. But art history alone quickly revealed itself to be a tool inadequate to my self-appointed challenge: to understand aesthetic experience. So I sought out insights from the history of technology, social theory, aesthetics, and even linguistics and literary theory.

All the while, I was delving into the writings and drawings of designers themselves, analyzing their styles and artistic visions, excavating the thoughts behind them. I learned of the French Enlightenment ideals underlying eighteenth-century neoclassicism; of the Neoplatonism and organic universalism propelling geometrically oriented designs from Andrea Palladio to Francesco Borromini to Anne Griswold Tyng; of the

late-nineteenth-century doctrine of structural rationalism, most powerfully theorized by E. E. Viollet-le-Duc, which influenced some early
modernists. I parsed the multifarious interpretations of functionalism,
from Ludwig Mies van der Rohe's "universal space" to Richard Neutra's
psychobiologism to Christopher Alexander's pastiche of sociology and
nostalgia in a "pattern language." Ideas garnered from all these practitioners and treatises and disciplines fed into an ever-evolving synthetic
framework that I was developing to shed light on how and why architects,
urbanists, and landscape architects design as they do, and on how people
experience the buildings and cities and places that architects design.

I learned a great deal about many things. But I was not satisfied and
was still searching for answers to how, and how much, the built environment affects what we think, feel, and do. Only creative writers, it seemed,
captured something of what I was trying to explain. The associative,
nonlinear, intuitive, and metaphorical thinking in poems and prose passages, some of which I include as epigraphs introducing each chapter,
crystallized some essential qualities of how people experience our built
environments. My initial questions still remained mainly unanswered.

Seven or eight years ago, I began coming across scattered writings—
in social theory, cognitive linguistics, various branches of psychology,
and cognitive neuroscience—that intimated a new account of how people actually perceive, think about, and ultimately experience their environments, which of course includes the built world. As I read more,
it became apparent that a newly developing paradigm, variously called
"embodied" or "grounded" or "situated" cognition, was emerging from
the confluence of work in many disciplines, some of them in the sciences. This paradigm holds that much of what and how people think
is a function of our living in the kinds of bodies we do. It reveals that
most—much more than we previously knew—of human thought is
neither logical nor linear, but associative and nonconscious. This still-
emerging paradigm provides the foundation for a model and analysis

of how we live simultaneously in this world, inside our own bodies with our feet on the ground; with other people; and in the worlds inside our heads, which are rife with simulations of the worlds we continuously imagine and reshape for ourselves. Human cognition, decision-making, and action are some admixture of all three.

This emerging, scientifically grounded paradigm of embodied cognition provides the foundation for the analysis I present here, which finally allows me to answer some of the questions that have occupied my time and preoccupied me for so long. In what ways, under what circumstances does a room or a building or a city square or any built environment affect us? What is it about a place that draws us in or repels us, that sticks in our memory or fails to register, that can move someone to tears or leave her cold?

While the seeds of the ideas I present here lay in that long-ago day in Florence, only after learning as much as I have of what scientists and psychologists currently understand about human cognition could I have written this book. I didn't get to the Piazza del Duomo by mistake; the design of the street I'd happened onto—the width of its sidewalks, the curving street that intimated glimpsing views of something large and white—nudged me along that path. Nor was I alone in the emotions I experienced in Florence's Cathedral Square. The plaza's strongly articulated boundaries, the scale and clear geometric forms of its central buildings, the sudden shift in their materials worked synergistically to capture my wayward attention. Its complement of scale-giving projections and recessions, its vibrant, colorful architectural detail—held my gaze by working with the operations and predispositions of the human body and mind to pack a powerful punch. That Giotto's bell tower and Brunelleschi's dome and the stately Baptistery have become a part of my autobiography—still alive, towering in my mind after so many years—is entirely explicable, a product of the nature and machinations of human long-term memory.

Think of how listening to your favorite piece of music can change your mood. How looking at an excellent painting transports you into another world. How an unusually shaped piece of furniture makes you muse upon the human body in repose; a dance performance activates thoughts of your own body in motion. Successful sculptures can incite imaginings of standing tall or slithering or floating; good film imbues our lives with story lines and dramas. Each of these arts affects us in ways that are powerful and real, but each does so only when we actively engage it. Usually that is only for a short time on any given day, and many days, not at all.

Our relationship to the built environment differs from that of any other art. It affects us all the time, not only when we choose to pay attention to it. What's more, the built environment shapes our lives and the choices we make in all the ways that these other arts do—combined. It affects our moods and emotions, our sense of our bodies in space and in motion. It profoundly shapes the narratives we tell ourselves and construct out of our daily lives.

What the new paradigm of embodied or situated cognition reveals is that the built environment and its design matters far, far more than anybody, even architects, ever thought that it did. The information this book reveals should have a profound impact on how people think about and designers construct the built environments of today and tomorrow. It holds a mirror up to show the worlds that we have made and clearly illustrates ways to remake our worlds to be less soul-deadening and more enlivening to human bodies and minds, communities, and polities.

Why do I feel so confident welcoming YOU to *your* world? Because in writing this book, I—who already brought considerable expertise about the built environment into the task—came to see buildings, landscapes, and cities afresh, and not just that memorable day in Florence, but every day.

Now I'm here to share what I've found with you.

The Next Environmental Revolution

... No world
wears as well as it should but, mortal or not,
a world has still to be built
because of what we can see from our windows ...
which is there regardless

—W. H. AUDEN, "Thanksgiving for a Habitat. I. Prologue: The Birth of Architecture"

Look around you. What do you see from your window? What do you see in the room surrounding you from where you sit? Your answer will be as singular as you are. Whether you live in a city large or small, in a suburb or exurb, or in one of the earth's growing megalopolises—New York, Seoul, São Paulo—you are in, and in all probability look out your window onto, a *built* environment.

In premodern societies, humans by and large constructed their own habitats which were dominated by the natural world. No longer. Around the world in developed and developing countries, people spend most of their time in and around buildings and constructed landscapes. In a hospital or bedroom you were born; in a hospice, bedroom, or hospital you will likely die. In houses and apartments you make your home. On streets and over bridges and underground you travel, going to the office or laboratory or manufacturing facility or the store where you work. In schools and community centers and playgrounds and parks you raise and educate your children, forge and nurture social bonds. In environments constructed for your leisure you go for walks, play ball or run, take in a theatrical performance, attend a sporting event, contemplate a museum exhibition, shop in stores, relax in cafés.

An ever-increasing percentage of humanity spends almost all its time—90 percent or more—inhabiting environments that have been conceived and constructed by human hands, and unlike in the premodern world, by hands that in most cases are not their own. Close to four billion people live in the most built-up portions of the world, known as urban areas. The places we inhabit and use in modern society are not only built in the sense that they are constructed, but are designed in the sense that they are constructed to look and function as they do because people made decisions. Someone *decided* to include one element or another, and someone chose to place these elements into a composition. Not just every building or urban square or park or playground—but every sidewalk is proportioned, every window is sized and positioned, because people made decisions about that sidewalk's dimensions and placement, about what weather that window would be able to withstand, or not, and about how it would look, feel, and function. Somebody decided. Somebody decided, whether that person thought much about the decision or not.

The built environment conveys a deceptive sense of permanence. In truth, statistics of change and growth show that it is constantly under renovation, renewal, and expansion. The numbers are daunting, yet only by confronting them can we comprehend what lies ahead. If we look toward just the next several decades, in the United States alone by 2050 the population is projected to increase another 68 million people, 21 percent, reaching almost 400 million. This will necessitate massive new construction in cities and the areas immediately surrounding them of buildings, landscapes, infrastructure, and urban areas. Think of increasing the built environment of the urban area of New York City by 20 percent—that would house 5 million more people. Then expand the urban area of Los Angeles to accommodate another 4 million people; of Washington, DC, to house an additional 2 million people; of Austin, another 400,000 on top of its 2 million people today, and so on, for every city in the United States. Consider all the construction needed

Construction site, United States

Construction site, China

in each of these cities to provide homes, office buildings, commercial areas, parks, and squares for that many more people. Housing them alone will require in aggregate perhaps a billion square feet of new construction. This, combined with an ever-changing economic landscape that draws people toward some cities and away from others, and which renders some kinds of buildings defunct and necessitates the reinvention of others, means that the United States will need many more, and many different kinds of, buildings and landscapes than the ones built by our parents and grandparents and forefathers and mothers. By 2030, as many as half the buildings that Americans will inhabit will have been built since 2006: every third, perhaps every second building you encounter will be new.

The construction campaign of the coming decades will be even more extensive outside the United States, so much so that it will make Americans' upcoming building needs look like a rounding error. Around the globe, a little more than half the world's population today resides in urban areas. Within less than two generations, urban growth in Asia, Africa, and Latin America will be so explosive that, by 2050, two out of every three people on the planet will be living in an urban area. That means 2.4 billion *more* people will need buildings in which to live, work, and be educated, as well as infrastructure through which to move, and landscapes in which to find refuge.

All this is beyond staggering.

Today 428 cities around the globe each house populations of between one and five million people. In the next fifteen years, that figure will increase to about 550 cities. There are currently 44 cities with a population of between five and ten million people; in fifteen years there will be 63. And the number of megalopolises, the gargantuan cities of more than ten million people, are expected to increase from 29 today to 41, also more than a 40 percent hike. Not only will urban populations grow at an unprecedented and breathtaking pace, the most complex

Site of new Ordos City, China, before (with artist Ai Weiwei)

Ordos, China

urban areas, the large cities and the megacities, will be at the forefront of the growth.

The human migrations to metropolitan regions and the construction campaigns that accompany all this are especially fervid in developing countries, notably in India and China, where by 2050 the number of urban dwellers is expected to increase by three hundred million to total more than one billion people. If China were building one entirely new city each year to house this population growth, it would construct a city the size of New York City every single year, for thirty-five years. In fact, in anticipation of this growth, China already has started and finished totally new, enormous cities in places where none had been before. To accommodate the vast migration of rural residents who will relocate to urban areas between now and 2030, China will eventually have 125 cities with a population of more than one million people. Its number of metropolises of five to ten million people will number sixteen, and it will have seven gargantuan cities of more than ten million people, with several of them greater than twenty million. This will require the construction of *four trillion* square meters of floor space in approximately five million buildings. In the next two decades, China will continue to expand its infrastructure—residential housing, roads, bridges, airports, power plants, water purification and distribution systems—on a phenomenal scale and pace that dwarfs anything humanity has ever seen. It will be equivalent in size to the existing urban infrastructure of the entire United States. Although China's rapid development seems to be in a class by itself, it really isn't. India's urban population is projected to increase during this period at an even more rapid rate, *doubling* in size, with an increase of four hundred million people. And other parts of Asia, and various countries in Latin America, are all growing and urbanizing at a fantastic rate, mostly in the form of slums and poor construction. Barring economic meltdown, this global construction tsunami will continue for decades to come.

Figuratively speaking, and almost literally speaking, all the world's a construction site. The decisions people make about the environments currently under construction, what and how we build and where, will affect the lives of billions of people for generations. If you consider all this a bit astonishing, look around you the next time you are out for a walk or a drive, and consider what it means that close to 80 percent of the places that Americans currently call home did not exist sixty years ago. Ask yourself what your life would be like, or your mother's or your brother's or your child's, if 80 percent of the buildings and streets and parks you see looked and were organized and functioned differently and better? What if all neighborhoods were vibrant and socially inviting? What if they all had convenient access to reliable, affordable, comfortable public transportation? What if everyone's house or apartment looked onto or was within walking distance of a well-designed and well-maintained park? If natural light streamed in through large, operable windows in every home, workplace, and classroom? Your life, and the lives of those you love, would be different—just as it would be different if you lived in a dark, characterless, cramped, windowless box, located somewhere in an undifferentiated thicket of high-rise towers.

Most of what we see from our windows or in our surroundings has been constructed, but it was not really *designed* in any but a rudimentary sense of the word. In the United States, 85 percent of new construction—whether it is a new bridge, an urban park, a housing development, or a school addition—is realized at the hands of construction firms collaborating with real estate developers or other private clients. Many of these builders bypass designers (a catchall term for professionals involved in designing the built environment, including architects, landscape architects, interior architects, urban designers, city planners, civil engineers, and other sorts of civil servants) completely, or employ them only cursorily, to review and stamp their approval on drawings—

drawings that have been prepared by people who all too often lack even basic professional training in design.

In the United States and in most other parts of the world today, many people believe that engaging a highly trained design professional is an unnecessary expense. True, wealthy individuals and corporations with plenty of assets do buy design to add beauty or prestige, and public and private institutions aspiring to serve as cultural stewards hire trained, informed professionals for complex structures such as skyscrapers. But this is not the norm.

The reason aside from financial considerations is that most projects in the built environment are commissioned on the basis of and judged by two complementary standards. Safety first: building codes and legislation and inspectors enforce standards that ensure that our bridges and buildings and parks and cityscapes will withstand gravity and wind, will weather the vicissitudes of climate and the ravages of time, and that their smaller features, such as electrical systems and stairways, will not shock or trip people up. Function next: people expect projects to serve an institution's or private individual's daily needs both effectively and efficiently, which often means with as little expenditure of resources—space, time, money—as possible.

Fair enough. People consider safety and functionality nonnegotiable. But the aesthetics of a new project, how it is composed, how the people who use it will experience it—how it is *designed*—is too often dismissed as unknowable or irrelevant. The question of how its design *affects* human beings is rarely asked, certainly not systematically, or centrally. People think that design makes something highfalutin, called architecture, and that architecture differs from building, just as surely as the Washington National Cathedral differs from the local community church.

This distinction between architecture and building—or more generally, between design and utility—couldn't be more wrong. More and more we are learning that the design of all our built environments matters so profoundly that safety and functionality must not be our only urgent priorities. All kinds of design elements influence people's experiences, not only of the environment but also of themselves. Good design—thoughtfully composed ordering systems and patterns, sensuously active materials and textures, deliberately constructed sequences of spaces—create coherent places that have a powerfully positive effect on people. Urban spaces, landscapes, and buildings—even small and modest ones—profoundly influence human lives. They shape our cognitions, emotions, and actions, and even powerfully influence our well-being. They actually help constitute our very sense of ourselves, our sense of identity.

We know that positive emotions prolong life and improve its quality. Yet too few people realize how extensive are the effects of design on human well-being and society's welfare. Rarely is design accorded the high priority that it deserves in our public policy and market calculations; rarely is it accounted for as we forever make and remake our worlds. Given the ongoing, literally world-shaping explosion in building across the globe, the time has come to confront a discomfiting truth. Our disregard for our built environments is bankrupting our lives. What's more, it threatens to bankrupt the lives of people for generations.

"Environment" is a word that sends most people's thoughts toward nature, and "environmental revolution" will elicit in most people's minds thoughts about overpopulation and pollution, particularly from carbon emissions, which has so degraded the protective ozone layer that encircles our planet that we face potentially catastrophic climate change. But the word "environment" refers simply to the places, circumstances, objects, or conditions that surround us. An environment can be ecological, social, virtual, or constructed. Its elements can be grass and trees, flesh and blood, words and images, paint and bytes, or it can be bricks, asphalt, and steel. And as we overwhelmingly live in environments given shape and dominated by brick, stone, wood and processed wood, glass, steel, and Sheetrock, it makes sense to deploy the word to describe the revolution we need to properly reconfigure our constructed world.

The environments we inhabit and build can make us and our children healthy or sick. They can make us and the people we love smart or dumb. Serene or despondent. Motivated or apathetic. What's more, it's their *design* that is in large measure responsible for these effects. A well-designed, properly constructed environment affects and supports our health, cognitions, and social relations. It meaningfully conveys to each of us that our human presence, not just our productive labor, credit card, or mortgage check, is valued. So how our buildings and landscapes and urban spaces are configured is not and cannot be only a matter of personal taste.

This book is a call to action, imploring all of us to do whatever it takes to develop a policy agenda and practical initiatives to better human welfare by improving the built environment. It is a call to all of us to develop, fund, and implement research programs that will expand our knowledge base about the ways we live and can live in buildings, landscapes, and cities. It is an exhortation to decision-makers in the private and public sectors to make a commitment to good design. And it is an exhortation to designers to devote their resources and attention to learning what is already known in other fields about the architecture of human experience.

Like a herald standing in the public square, I am trumpeting the day's news, and here is my plea. Listen. The shape of our built environments is largely driven by interests, including but not confined to the marketplace, in which many people make decisions that are not necessarily in society's or the planet's best interests. If we've learned anything in the past twenty years, it's that people are not rational actors—at least not most of the time. People think about and experience their built environments in ways that comport with what we know and are learning about human cognition, social behavior, and experience more generally. And the places in which most of us live are in one way or another, in many ways or just a few ways, not the places we need. This holds true for our internal, individual experience and for the way we conduct ourselves as members of groups in society.

In chapter 1 we assess the actually existing contemporary built environment, taking stock of how it does and mostly does not accommodate humans as we know them to think, feel, and act. Chapter 2 lays out the foundation of how people experience their built environments by looking at ordinary cityscapes and civic landmarks, explaining the nonconscious cognitive mechanisms underlying our human experience of the constructed world, which powerfully shape our thoughts as well as our individual and social lives. The next three chapters examine a wide range of buildings, landscapes, and urban environments to excavate the specific dimensions of how our experience of the built world is determined, shaped, and inflected by the fact that we are situated: in the body (chapter 3), in the natural world (chapter 4), and in the social world (chapter 5). Chapter 6 puts all that we've learned together to advance some basic principles for how built environments can be designed for the humans that people are. In the final chapter, we discuss the larger implications of our findings, how all this establishes the absolute centrality of built environmental design to human well-being, now and for the future.

What kind of worlds and societies do we want to shape for the generations to come? This remains as pressing a question today as it was in 1943, when Winston Churchill, following the Germans' destruction of the chamber of the House of Commons in London's Houses of Parliament, urged Britain's parliamentarians to vote to rebuild the chamber in its original rectangular form, with two long rows of benches facing one another, accommodating and at the same time symbolizing

The House of Commons after the London Blitz

the two opposing parties. The two-party system that this arrangement represented, Churchill maintained, constituted the backbone of British parliamentary democracy. Emphasizing how design shapes everyday experience, Churchill declared that "we shape our buildings, thereafter they shape us."

The importance of Churchill's pronouncement basically has been overlooked. The built environment per se remains, for the most part, little discussed. The media cover some aspects of it, but mainly in the context of "starchitecture," travel destinations, or home decor. In the meantime, the amazing breakthroughs in cognitive neuroscience and perception are establishing precisely why our relationship to the built environment is so essential to the human experience, and describing how.

Certainly, some writers have led the way in considering how the built environment's design overtly and subtly shapes the type and character of people's social interactions. Jane Jacobs's *The Death and Life of Great American Cities*, published in 1961, launched a broad attack on early postwar American city planning's slum-clearance and development policies, suggesting that even well-intentioned interventions could profoundly compromise people's lives. Jacobs argued that the forms of our cities and urban places must be based on empirical knowledge about how urban dwellers actually conduct their social and individual lives, a position she learned from urbanist William H. Whyte, who spent decades studying people in public spaces, examining the design elements that attract or repel passersby. A decade after the publication of Jacobs's book, Oscar Newman, in *Defensible Space*, substantiated Jacobs's assertions by linking the incidence of crime to the design of just the sort of social housing projects Jacobs had criticized. Newman identified design elements, such as homogeneity, repetition, and the absence of sight lines, that prevented residents from surveying and developing a sense of emotional connection to the places they inhabited, thereby eroding their ability to develop a robust feeling of responsibility for their community. Recently, the influential Dutch urbanist Jan Gehl extended the work of Jacobs, Whyte, and Newman, specifying the design elements that contribute to vibrant urban settings, such as "soft" edges, walkability, active ground-floor spaces, and variability.

Jacobs, Whyte, Newman, and Gehl's works demonstrate the thoroughgoing effects of design on people's social lives. By contrast, analyzing how the built environment shapes and impacts people's *individual* experience has been mostly relegated to the province of theoretical and philosophical speculation, in works such as Gaston Bachelard's *The Poetics of Space* and Edward Casey's *Getting Back into Place*. The most notable exception, an empirical study by Kevin Lynch (*The Image of the City*, 1960), is more than fifty years old. Lynch conducted interviews with urban dwellers and drew on principles from Gestalt psychology to construct an intuitive framework of how urban dwellers make sense of a city and where they find themselves in it. He discovered that people, to navigate complex environments and develop an internal cognitive map of a city's organization, rely on very specific design elements, a combination of *landmarks* (the Eiffel Tower); *edges*, which must be clearly defined by visible boundaries (the facade lines of Parisian boulevards); and demarcated *paths* that link to focal points, or *nodes*, such as plazas, squares, and major intersections.

Of all these studies on how we as individuals experience built environments, only the findings of Kevin Lynch have been substantively confirmed: landmarks, edges, paths, and nodes are indeed the critical tools our brains use in human spatial navigation and cognitive mapping. Recently, a group of cognitive neuroscientists—Edvard Moser, May-Britt Moser, and John O'Keefe—reinterpreted and further specified Lynch's paths and nodes. In discoveries that collectively earned them a Nobel Prize in Physiology and

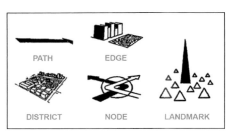

Paths, edges, nodes, landmarks: the central tenets of wayfinding, from Kevin Lynch's *Image of the City.*

Medicine, they identified specific place recognition and building recognition cells, which work together with grid cells in an integrated system. An inner GPS in our brains enables us to orient ourselves in space. Now we know the answers to questions such as "How do we know where we are? How can we find the way from one place to another? And how can we store that information so that we can retrace the same path another time?"

Lynch's work highlighted the need for more information on the ways that humans experience and are impacted by their built environments. And in small, relatively confined corners of the academy, research did continue. But few findings reached the eyes, ears, and minds of the people who buy and live in buildings—clients—or even designers. More and more, such questions are being taken up in various research initiatives and collaborations among urban, architectural, and interior designers and researchers in the academy and the health-care industry, as well as being championed by the small but growing Academy of Neuroscience for Architecture.

To investigate how our built environments shape both our internal and our external worlds—in common parlance, how we *experience* them—we must first articulate what we mean by an "experience." An experience differs from the unselfconscious fact of mere existence; it is distinguished by its unifying quality, which pervades all its constituent features and gives them meaning. This persuasive unity is the product of the human mind, through which everything we encounter is filtered and interpreted.

In the past two decades, a vast amount of knowledge about the operations of the mind—much of which is not focused on architecture and the built environment per se—has emerged from the sciences and social sciences. Synthesizing this knowledge produces a surprising yet unavoidable insight: our built environments will not accommodate people's needs until we integrate what we know and are learning about human experience into their design and composition. This is true at every level, from families inhabiting their homes, to schoolchildren on playgrounds, to workers toiling in the offices or distribution centers of corporations.

This steady stream of new research on how humans perceive and think collectively demonstrates that humans are pervasively integrated into their environments. Whether it's the way we see, without really knowing that we see, how lines arrange into patterns on a wall, or nonconsciously register the height or shape of a ceiling, or respond without realizing it to the quality and intensity of light in a room; whether our intuitive sense of gravity has been tamed or tempted, how we imagine feeling the coldness of a stone floor—a person's sense of emotional well-being, her social interactions, and even her physical health are all affected by the places she inhabits, in ways large and small. This fast-growing body of knowledge originated in the "cognitive" turn in psychology in the 1960s, when ever more scientists began to argue that people's thought processes—their cognitions—could be scientifically studied and were as important a dimension of human experience as human behavior. The cognitive revolution continued to gain speed, then accelerated rapidly in the 1990s, when a number of new imaging and computation technologies permitted the scientific study of the human brain in action.

We know much, much more—a hundred times more—than we did a few decades ago about how cognitions directly or indirectly affect, or are precipitated by, our experience of the built environment. And we now know that, even if some of what passed for conventional and scientific wisdom for centuries about what makes good architecture, landscapes, or urban design was right, much more of it was and remains just plain wrong. What we know about the structure of human memory, learning, and the relation of emotions to cognitions has been utterly transformed. Not only do we understand the mechanisms of spatial navigation, thanks to Lynch and his successors, but we are finding also that those mechanisms play important roles in other cognitive processes that are essential to our daily lives. We know that our perceptions and determinations to act do not happen entirely sequentially but instead are more intermeshed. And most important, we know that the bulk of our cognitions are nonconscious and associative in nature.

We need a new conceptual framework to understand how we think about and experience the built world, because the human brain fundamentally differs from the brain that psychologists, philosophers, and designers until fairly recently thought we had. During my childhood, a cardinal belief among psychologists was that the physiology of the human brain, after a critical period early in a person's development, was set. No new neurons could be generated, no new connections established or pared away. Then around 2000, a series of studies of London taxi drivers demonstrated clear changes in cabbies' brains (in particular, their hippocampi) after they completed the extensive training required of them to memorize—in Lynch's terms, to build a cognitive map of—the city's geography. Even in fully formed adults, these and other studies revealed, the human brain is dynamic, ever-changing in response to what we experience in our environments—human, social, physical, architectural, landscape, and urban. The fact of our brain's neural plasticity has immense implications for our understanding of human cognition: it reveals that as we learn, our brains are changing shape, rewiring us throughout our lives. And contrary to what had been assumed for centuries, we now know that all our lives, our minds are changing and *quite literally being shaped* by our experiences in the physical environments in which we live.

The more we know, the more we can think about, investigate, and assess the fit between what we have built and will build and what it takes to nourish our well-being. The more we know, the more evident it becomes that we must revisit our received wisdom about cities, architecture, landscape architecture, and the built environment's relationship to people. And we should undertake this reinvestigation with optimistic dedication, with hopeful vigor. My decades of studying and writing about the built environment have made it clear that our built environments can be made much, much better. At every level of investment, there is much that all of us can do to improve our buildings, landscapes, and cities. And it turns out that, more often than not, it takes just as many resources to build a bad building—or landscape or townscape—as a good one.

So consider the shape of the room where you are sitting and the height, shape, and color of its ceiling. The texture and construction of the walls.

The softness or hardness of the floor surfaces. The views to nearby internal spaces and the views (if you have them) through windows to the outdoors. The air quality and temperature. The quality of the sounds that you hear. The selection of the furniture and its arrangement. The types and levels of lighting. The configuration of the passageways leading to nearby rooms and places, and their placement relative to where you are. All of this affects you. It affects your well-being and your health, in ways you may recognize and in other ways you may not even suspect. It affects the ways you interact with—and even conceive of—the other people in that space. It can affect your very sense of who you are, as someone who belongs or does not belong in that type of place.

Why does this matter? Because it can be changed. Everything around you—from the shape of the room in which you currently sit, to the amount of sunshine filling your home, to the character of the house or apartment where you live, to the width and patterns on the sidewalks or roads that brought you there—is as it is because somebody made a *choice*. By commission or by default, the built environment is composed, which means that it could have been composed differently. And much of it can be remade, as so much more of it will be created in the coming decades. We have before us an unprecedented opportunity to reshape the world into a better place.

Louis Kahn, an American architect who built some of the most revered buildings of the late twentieth century, spent his life making the case for the powerful effect of built environmental design on people's lives. Once he put it this way: "If you look at the Baths of Caracalla . . . we all know that we can bathe just as well under an eight-foot ceiling as we can under a 150-foot ceiling." But, he insisted, "there's something about a 150-foot ceiling that makes a man a different kind of man." In

"There's something about a 150-foot ceiling that makes a man a different kind of man," said Louis Kahn about the Baths of Caracalla (reconstruction).

speaking of the ennobling quality of the legendary Roman baths, Kahn expressed an intuition that has been proved correct, albeit for reasons that Kahn himself could not possibly have known. A recent study revealed that people think more creatively and respond better to abstract concepts when seated in rooms with high ceilings. A person who feels quite literally "unconstrained" is more apt to think creatively.

Architecture has always seemed to me to be the most important art—the kind that everyone deserves. Our buildings, landscapes, and cityscapes influence the lives not just of the people who commission or pay for them—and if they are constructed for investment, as most are, not ever inhabited by them. They are foisted upon countless users and passersby. Moreover, most buildings, landscapes, and urban areas outlast people, not only those for whom they were originally constructed but also subsequent generations of people, and sometimes even beyond.

Of course some people, especially in the design professions, know that design matters. But many find themselves stumped when making the case for *why* design matters, and matters crucially to people's lives. One such person I know runs a small, successful nonprofit that advocates for good architecture. She once told me that when a new crisis that affects the design world hits, like the collapsing levees in New Orleans, the prospective demolition of a landmark building, or the permitting of a new real estate development sure to be execrable, she sits around a table with members of her board bemoaning the state of the built environment. These are all like-minded professionals, all passionate about design. We all keep saying to one another, she once complained, that design *matters*. But no one, she continued, can ever say very much about why.

Before, perhaps, no one could. Now we can.

WELCOME TO YOUR WORLD

CHAPTER 1

The Sorry Places We Live

Boredom sets in first, and then despair.
One tries to brush it off. It only grows.
Something about the silence of the square.
— MARK STRAND, "Two de Chiricos. 2. The Disquieting Muses"

The fact that a man does not realize the harmfulness of a product or a design element in his surroundings does not mean that it is harmless.
— RICHARD NEUTRA, *Survival Through Design*

The headline to one of my early essays in architecture criticism, published in *The American Prospect,* was written by my editor, not by me. But when he sent along the galleys, it was clear he thought I'd pulled no punches: "Boring Buildings," the title thundered. In the subhead I could almost hear my own plaintive voice, wondering *"Why Is American Architecture So Bad?"* Since that essay's appearance fifteen years ago, our country's and our world's political, social, and economic landscapes have much changed. The events of 9/11 ushered in a far more perilous and increasingly self-conscious era. The Internet and digital technology changed how we communicate and how we shop, the nature of our right to privacy and even our sense of ourselves; they also rapidly accelerated economic integration and made globalization an economic, social, and cultural reality. Even so, fifteen years later, the indictment announced in that headline rings true, and not only in the United States.

Four Sorry Places

In the places we live, beggary casts a wide net, as four very different kinds of settings exemplify. The shacks inhabited by many millions of people on every continent except Antarctica make the case that built environments lacking in any sort of considered design greatly contribute to the degradation of human life. If we consider such slum dwellings in relation to the developer-built single-family houses that hundreds of millions more call home, however, we see that lack of resources constitutes only one part of the problem. Then, if we examine both alongside the design of a resource-rich high school in New York City, it becomes clear that people's failure to accord their built environments sufficient priority plays an important role. And if we take all that information and synthesize it with what we can learn from an art pavilion in London designed by a Pritzker Prize–winning architect, Jean Nouvel, we see that even with ample resources, good intentions, and well-placed priorities, things go wrong. These four examples of built environmental beggary show how pervasively poor our built environments are, and suggest the complex reasons why. They also belie what slum dwellings by themselves might suggest: money, or lack of it, constitutes only a small part of the challenge.

Slums are an example where resources truly are scarce. In Haiti, one and a half million people lost their homes in the devastating 2010 earthquake, and many lived in the aftermath and continue to live in encampments of makeshift, temporary shelters. Among the thousands of heartbreaking images documenting the Haiti disaster was this photograph of a huddling row of shanties clutching to the median strip of Route des Rails in Port-au-Prince, where entire families inhabit tarp-covered one-room shacks with dirt floors. Cars and trucks rumble and speed by. No electricity. No plumbing. No privacy. No quiet. No clean air to breathe, fresh water to drink. Just unlucky

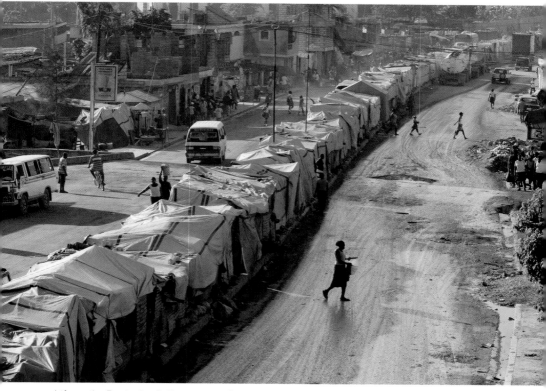

Life on the Route des Rails one year after the catastrophic 2010 earthquake, Port-au-Prince, Haiti

people trying to maintain their rectitude and dignity in a built environment that pulls them down every day.

Although this photograph depicts people struggling to survive in the wake of a singularly catastrophic event, the circumstances in which they live are unusual in only two ways: cars pass through this linear settlement very quickly, and fabric tarps cover their shacks instead of the usual corrugated metal, scraps of plastic, thatch, or rotting plywood or cardboard sheets. These Haitian shacks otherwise resemble their analogues: favelas in Brazil, bidonvilles in French-speaking countries such as Tunisia, townships in South Africa, shantytowns in Jamaica and Pakistan, *campamentos* in Chile, and to use the most generic name among the dozens of such appellations in usage, slums around the world. The names, the materials from which they are built, the destitution of shelter they provide, the degree of desperation within differs depending upon economy, culture, climate, and continent. But the basic living arrangements are the same. One, two, even three generations jammed together with their belongings in one or two insalubrious rooms that lack the basic infrastructure of power and sanitation.

Thirty percent of all South Asians, including 50 to 60 percent of the residents of India's two largest cities, Mumbai and Delhi, live in slums; estimates of the population density of Mumbai's Dharavi slum range widely, from between 380,000 and 1.3 million people per square mile—five to nineteen times that of Manhattan. Sixty percent of sub-Saharan Africans inhabit slums. Four million people dwell in the largest slum in the world, draped across the outskirts of Mexico City. All told, one out of every seven of the planet's people, totaling one billion, and one-third of all urban dwellers call such places home. The Housing and Slum Upgrading Branch of UN-Habitat predicts that by 2030 the number of people living in slums will more than double, as slums are "the world's fastest-growing habitat."

How might a child growing up in a leaky shack in Port-au-Prince or Mumbai or Lagos be affected by the physical circumstances of his surroundings? Children who inhabit chaotic, densely populated homes exhibit measurably slower overall development than do children raised in more spacious quarters. They underperform in school and exhibit more behavioral problems both in school and at home. An acoustically uncontrolled space too tightly packed with people, with little privacy, correlates with disorder, explaining why crowded homes are associated with higher rates of child psychiatric and psychological illness. We know that the

Slum dwelling in Africa

degree of control a child feels he has over his home environment is inversely related to the number of people per square foot who live there, and a diminished sense of control compromises a child's sense of safety and autonomy, of agency and efficiency, and hence likely of motivation.

This is just the most obvious way that the design of a shanty house compromises people's lives. Overcrowding, lack of privacy, and environmental noise diminish a child's capacity to manage her emotions and hinder her ability to deal effectively or even to cope with life's challenges. So not only do slum-dwelling children enjoy fewer opportunities, but they are also less capable of taking advantage of the opportunities available to them. Even if an adult raised from birth on the Route des Rails unexpectedly encountered sudden good fortune, she would likely struggle, and struggle more than a woman whose childhood had not been permeated with such built environmental deprivation and degradation. A person's experience of growing up in challenging and impoverished circumstances results in *lifelong* diminished capabilities.

It's not only such patently deprived places, lacking in any sort of design, that diminish people's well-being. Resource-rich middle- and upper-middle-class housing developments in the United States prove that. Consider two new ones that differ greatly in locale, price point, consumer base, and design. Lakewood Springs is in Plano, Illinois, a suburb approximately one hour's drive west of Chicago. A relatively small middle-class development, its homes are arrayed on the paper-flat land that constitutes much of America's midwestern plains. Stylistically, the architecture at Lakewood Springs reinterprets the traditional midwestern farmhouse in two basic models, with low-slung, single-story homes and two-story townhouses alternating along cul-de-sacs and gently curving streets. The second settlement can be found in Needham, Massachusetts, a development of McMansions. Homes here run larger, and though their configurations vary more widely, they hew stylistically to what might be termed Realtor Historicism.

Middle-class developer-built surburban home

Size and price differences notwithstanding, the $200,000 house in Plano, and the $1 million house in Needham, share many basic features and problems. Each unit accommodates a single nuclear family. In spite of clear demographic trends toward the aging of our population, in spite of the growing variability of family composition, there is no place at Lakewood Springs or in Needham for elderly parents or for the disabled incapable of living on their own. Within each development, lot sizes are more or less identical (bigger in Needham), with houses plunked down at the lot's midpoint, sandwiched between a front and back yard. Residents enter their homes through the garage, yet "front" doors gaze sorrowfully onto the street while "front" yards go largely unused. The layout of the homes and the scarcity of locally accessible services afford inhabitants limited opportunities for spontaneous social interaction.

Both in Plano and in Needham, these homes are tossed together from off-the-shelf materials using simple construction techniques

Higher-end developer-built surburban home

requiring little skilled labor. Their environmentally suspect materials are cheap and thin; their timber harvested with little regard for sustainability; the PVC piping threaded through them leaching volatile organic compounds into the ground and the water the inhabitants drink; and gypsum walls that visually demarcate one room from the next offer but scant acoustic or thermal insulation. Carelessly standardized room arrangements and stock floor plans result in poorly placed windows and rooms with little attention to where the house happened to end up on the lot. No heed is paid to prevailing winds or to the trajectory of the sun's rays. For example, in one house, the living room might be dark, while in another, it might blaze with sunlight; some bedrooms might be too cold, others too warm. Efficient thermostats and artificial light cloak poor design.

Okay, you may think, so these middle- and upper-middle-class housing developments aren't great. Surely the homes and institutions and landscapes that serve more affluent people are better? Some are, but many more are not. Take my own experience with trying to find a school for our son. A few years ago, preparing to move to New York City from out of state, my family visited a number of private schools in Manhattan and Brooklyn in search for the right fit for our soon-to-be high schooler's challenges and considerable abilities. Most families would be happy to send their children to a certain school in upper Manhattan. This pre-K–12 school occupies a collection of buildings abutting one another on a leafy side street; its entrance sits in a heavily shadowed, deeply sculpted, Richardsonian Romanesque masonry pile of earthy tans and reddish browns. But its high school, located in a newer building perhaps forty years old, looked a little less and a little more like a hundred suburban high school degree factories. Its many classrooms, below-grade, were rectangular cinder-block grottoes, with industrial-grade wall-to-wall carpeted floors and ceilings lined with standard-issue white acoustic tiles. Furniture for the classroom consisted of metal desks and chairs. Ninth-, tenth-, and eleventh-grade classrooms lined a narrow linoleum-tiled internal corridor, with sound bouncing around and off walls like so many balls on a crowded playground.

It got worse. Even though one of an adolescent's principal challenges is to learn how to navigate an increasingly complex social world, only one large space explicitly facilitated this important pursuit, an informal gathering area that students called the Swamp, a nickname that conveys, presumably, not the room's physical appearance but a student's experience there. Forlorn castaway sofas floated in a cramped afterthought of a corridor, an unwelcoming muskeg that offered a prix fixe menu of social opportunities: being or not being part of a large, amorphous group. In the Swamp, teenagers buzzed like dragonflies and crickets and locusts, a deafening din.

Yet research clearly demonstrates that design is central to effective learning environments. One recent study of the learning progress of 751 pupils in classrooms in thirty-four different British schools identified six design parameters—color, choice, complexity, flexibility, light, and connectivity—that significantly affect learning, and demonstrated that on average, built environmental factors impact a student's learning progress by an astonishing *25 percent*. The difference in learning between a student in the best-designed classroom and one in the worst-designed classroom was equal to the progress that a typical student makes over *an entire academic year*. Students participate less and learn less in classrooms outfitted with direct overhead lighting, linoleum floors, and plastic or metal chairs than they do in "soft" classrooms outfitted with curtains, task lighting, and cushioned furniture, all of which convey a quasi-domestic sensibility of relaxed safety and acceptance. Light, especially natural light, also improves children's academic performance: when classrooms are well lit—and most especially when they are naturally lit—students attend school more regularly, exhibit fewer behavioral problems, and earn better grades. Windowless rooms of the kind in the high school we visited exacerbate children's behavioral problems and aggressive tendencies, whereas daylit, naturally ventilated classrooms contribute to social harmony and facilitate good learning practices. And the sort of noise that we heard that day detrimentally impacts learning, just as it does children's sense of well-being at home, communicating to inhabitants their lack of control over their surroundings. This in turn elevates their stress levels, further inhibiting their learning.

Why did this school and why do so many of this country's schools construct these kinds of inadequate buildings for their high schoolers? Why, today, in light of readily available research demonstrating that the design of learning environments can inhibit or advance pedagogical objectives, do they continue to use them? The particular high school we visited that day is neither a for-profit institution nor a cash-starved, high-aspirational nonprofit. The reason the school's board of trustees didn't consider it a top priority to create a physical plant that supports students and facilitates

A school building that helps students learn: Sidwell Friends Middle School (KieranTimberlake), Washington, DC

their transition to college and emerging adulthood was not so much a lack of concern or resources but a lack of awareness.

What about when clients do recognize design's importance, and are willing to devote ample funds to it? Even then things go wrong. Consider one of the pavilions built by London's popular Serpentine Galleries, which annually selects an internationally celebrated architect to design a temporary pavilion in Hyde Park. Crowds throng the annual Serpentine Pavilion; many more who don't make it to London see it in the style pages, or online. In 2010, the globe-trotting French starchitect

Jean Nouvel took the Serpentine commission's impermanence and flexibility as license to construct a luxuriously finished, angularly diagonal cacophony of steel, glass, rubber, and canvas, with blazing transparent, translucent, reflective, and opaque red surfaces. Explaining that he wanted to evoke the image and sensations of a setting summer sun, Nouvel romantically recounted his "building from a dream," an expressive place to "catch and filter emotions, to be a little place of warmth and delight" where people can have "some, I hope, happy sensations."

Nouvel's pavilion looks terrific in photographs, like a red thunderbolt shot through the verdure of Hyde Park. But I doubt that the people who visited it lingered there to play on its bloodred oversize chessboard, and I can nearly guarantee that fewer still experienced the happy warmth and delight that Nouvel aspired to elicit. That is because his three most significant design decisions—sharp angles, diagonally canted walls and ceilings, and the large-scale use of the color red, including red-tinted glass for the windows—were guaranteed to undermine rather than advance his intentions. Instead of "happy sensations," it's far likelier that visitors to that year's Serpentine Pavilion felt an uneasy sense of stress, even anxiety.

Humans respond to compositions dominated by sharp, irregular, angled forms with discomfort, even muffled fear. The color red and red light stimulates people, but generally not in a pleasant way. We know that when infants or people suffering from mental illness are exposed to red light, anger and anxiety are heightened. Normally functioning adults, when placed in red rooms or exposed to intense red light, suffer a diminishment of problem-solving and decision-making abilities and a reduced capacity to engage in productive social conversation. An all-red environment shifts the human pituitary gland into high gear, raising blood pressure and pulse rate, increasing muscular tension, and stimulating sweat glands. Such a place can energize and excite us, to be sure, but it's the kind of excitement that's coupled with agitated tension and can easily slip into anger and aggression.

2010 Serpentine Pavilion (Jean Nouvel), London, England [demolished]

Imagine being inside Nouvel's pavilion, then compare that experience to standing next to or inside a different temporary exhibition pavilion, Thomas Heatherwick's Seed Cathedral, constructed the same year as Nouvel's to represent the UK in the World Expo held in Shanghai. Built on a modest budget, Seed Cathedral's dandelion-like structure was set into a silvery, Astroturf-covered landscape. Protruding from the inside and the outside of a simple wooden box were 60,000 clear acrylic rods, their tips embedded with one or several seeds—a total of 250,000 of them, harvested from Kew Gardens' Millennium Seed Bank. The Seed Cathedral's evenly spaced, pappus-like transparent rods captured the light of the sun and transmitted it into the pavilion's interior. Each individual rod also held a tiny light source, so that at night, the feathery Seed Cathedral displayed literally 60,000 points of light, softly swaying in the wind.

Same year, two pavilions, totally different experiences: Nouvel's excited agitation, Heatherwick's inspired gentle delight. The comparison starkly illustrates how much our thoughts and even our moods can be influenced by the design of the environments we inhabit when the environments are at the extreme. Yet this is also, and profoundly, true for *ordinary and conventional* design.

These accounts of shacks in slums, suburban developments, a New York City high school, and London's 2010 Serpentine Pavilion all exemplify failures of the built environment. Such failures happen far, far more often than not. The example from Haiti shows what happens when grossly insufficient resources are devoted to the places we live: such places are all but doomed to be wholly inadequate, indeed harmful to their occupants. But the next three examples—the suburban housing developments, the high school, and the elite art pavilion—complicate the story considerably. Individually and collectively, they demonstrate that even when people devote resources to design and construction, things can still go very wrong.

Why have we been condemned to live in mostly boring buildings, to spending a good deal of our lives in landscapes and places distinguished mainly by their beggary?

The answer is not primarily resources. At virtually any level of resource allocation, as measured by cost, design can be poor or better, with better design enriching buildings, landscapes, and urban areas. As we will see, even housing for the world's poorest can be better than the slums and favelas in which a billion people live. And plenty of resources were channeled into the Plano and Needham developments. The problem was an overly conventional and narrowly bottom-line approach to decisions about the design of their buildings, infrastructure,

Seed Cathedral, 2010 World Expo Pavilion (Thomas Heatherwick), Shanghai, China [demolished]

Seed Cathedral, detail showing seedpods encased in resin rods

and landscapes, resulting in stultifying environments that functioned far less optimally—or to put it negatively, far more harmfully—to the well-being of its inhabitants than they could have. The New York City school's trustees lacked neither resources nor good intentions, but they likely wanted for knowledge about the multiple profound effects of design on students' cognitions and emotions, learning and achievement, individual well-being and communal cohesion. Yet as the 2010 Serpentine Pavilion illustrates, even when educated clients with deep pockets hire truly gifted designers, success is by no means guaranteed.

Why do developers build single-family houses that are patently wasteful and do not serve contemporary living patterns? Why did the board of trustees of an elite high school use for so many years a building that failed to promote learning—in fact, very likely inhibited it? Why did Nouvel, an internationally acclaimed leader in the field and an architect who has some excellent buildings to his name, fail to create the "happy" pavilion he believed he had designed in Hyde Park?

The problem is an information deficit. If people understand just how much design matters, they'd care. And if they cared more, they'd change.

A Bird's-Eye View of Built Environmental Beggary

Let's zoom away from these initial four examples to examine our landscapes around the globe with fresh eyes. In what ways are, and are they not, the ones that people need? From low building to high architecture, from metropolitan planning to suburban design, from streetscapes to landscapes, the examples discussed above represent a much larger, indeed all-too-common reality. Even in resource-rich countries and cities and institutions, people live out their lives in misconceived, poorly designed, and badly executed places. Billions suffer the consequences, mostly clueless that the origin of many of their social, cognitive, and emotional problems may lay directly beneath their feet and above their heads.

Discard the notion that design is a discretionary luxury. The built environment affects our physical health and our mental health. It affects our cognitive capabilities. And it affects the ways we form and sustain communities. The built environment affects each of these facets of our lives, and because they are related to one another, it does so in ways that are mutually reinforcing. Our asphalt-draped cities starve us out of a healthy ongoing relationship with nature. Wherever we look—at infrastructure or urban areas or suburban settlements; at landscapes or cityscapes or individual buildings, the bottom line is boring buildings, banal places, and hoary landscapes.

The Infrastructure of Urban Life

Infrastructure provides the cylinders in any country's engines of economic growth, yet in developing countries and many wealthy countries—and in the United States—infrastructural design and maintenance is in a deplorable state, revealing how little our society values even those parts of the built environment that are essential to economic growth. We traverse crumbling bridges and corroded, eroding pipes that occasionally blast their way into the public awareness. In the United States in 2013, according to the American Society of Civil Engineers (ASCE), one infrastructural system after another fails us with insufficient capacity, deplorable condition, and more. The ASCE scores our society's most essential structures and systems on the same scale every schoolchild knows. Our "report card" is shameful: public transportation systems earned a D+, roads, a D, and public parks and recreation facilities, a C+. Decades of neglect meant that in 2013, to bring our infrastructures up to *minimum* standards, *$3.6 trillion* in the next seven years was needed. While the infrastructure of Japan and much of

Western Europe is in better shape, many more countries in Africa, Latin America, and parts of Asia suffer with infrastructures that remain poor or effectively nonexistent. In India the situation is so extreme that in some places, private citizens have taken to hiring their own contractors to construct roads, sanitation systems, even bridges, resulting in a piecemeal infrastructure that serves mainly those who pay. In much of Africa and Latin America, infrastructure for water distribution, sanitation, transportation, and information technology simply does not exist, and in Latin America, recent reports suggest that public investment in these systems appears to be diminishing.

The character of cities in the developed world—those of rapidly growing Asia, those of increasingly urbanized Latin America and Africa—differ from one another markedly. Even within these vast regions, variations are enormous: just as Washington, DC, differs from New York City, Beijing is quite unlike Shanghai, let alone Mumbai, Lagos, Los Angeles, and São Paulo. But whatever their variations, many

Faulty design can kill: in 2007, the I-35W bridge in Minneapolis collapsed during rush hour, killing 13 and injuring 145 people.

cities around the globe suffer from the impoverishment of thought-lessly created urban spaces. The "green" spaces they lack or pretend to provide, the materials and quality of construction of their buildings and landscapes, the relationship of their buildings to their sites—all fail to promote and many actively undermine the welfare of their citizenry, individually and collectively.

Humans crave and need access to the outdoors and to nature and suffer in its absence, yet few of us appreciate how fundamental that need is. Contact with nature confers on people salutary effects that are nearly immediate. Twenty seconds of exposure to a natural landscape can be enough to settle a person's elevated heart rate. Just three to five minutes will suffice to bring high blood pressure levels down. Nature, quite literally, heals us: hospital patients recovering from gallbladder surgery, when placed in rooms with views of deciduous trees instead of rooms facing a brick wall, healed so much more quickly that they were released from the hospital *nearly a full day earlier.* Even during their hos-pital stay, they felt less perceived pain as measured by how frequently they requested pain medication.

Not surprisingly, then, people's preferences for urban environments also consistently skew toward nature, and cities containing ample green space consistently rank higher on lists of desirable places to live. When people in one large study were asked to prioritize a list of features that they would use to determine a neighborhood's desirability, "access to nature" consistently ranked first or second. Merely the view of grass and other greenery from a residential window increases the likelihood that an inhabitant will be satisfied with his neighborhood. And still, in spite of it all, many of the world's major cities contain, shockingly, less than 10 percent green space, including Bogotá (4 percent); Bue-nos Aires (8.9 percent); Istanbul (1.5 percent); Los Angeles (6.7 percent); Mumbai (2.5 percent); Paris (9.4 percent); Seoul (2.3 percent); Shanghai (2.6 percent); and Tokyo (3.4 percent). When it comes to the inclusion of nature in cities, political will emerges as a determining factor: many of the cities containing 35 percent or more green space are located in countries where the government takes a strong hand in managing public resources and is devoted to the public welfare: London, Singapore, Stockholm, and Sydney.

From the point of view of public health and human welfare, many of the world's cities—and suburbs—are wanting not only in their open spaces but in the poor quality of their construction and the cheapness of their materials. After every natural disaster we learn whether or not a given city is riddled with poor quality concrete that crumbles and cracks soon after the mixing trucks disappear. Suburbs and cities are laced with low-grade timber and shoddily manufactured, toxin-emitting composite woods. Everywhere, cloddishly designed mullions and moldings knit together haphazardly, inexpensive plastics—this is the stuff of which our urban environments are mostly made. Cheap construction disintegrates. Cheap buildings make cities that cheapen lives.

Even though we know more than enough about human physiological and psychological needs to know that buildings and cities must accommodate more than basic functional necessities such as sleeping and eating, they must also nurture our social connections and sense of belonging to a community. And even though we know that communities gain character through their interaction with distinctive, geographically identifiable places, woeful inattention is paid to such features of human social interaction, or to how buildings relate to their sites. Workplaces all too often neglect people's need for privacy. Although load-bearing brick makes sense for buildings in rainy, muddy climates but less so in drought-prone regions, all too often the design of buildings from these different climates is more or less identical. Contractors, instead of harvesting materials locally, ship them in bulk, transcontinentally. Local construction practices are neglected or deliberately forsaken. This lack

of attention to local climates, cultures, materials, and construction practices is owing, in many places, to much too much building too quickly,
as is the case in China. In other places, such as the cities of sub-Saharan
Africa, it is owing to an absence of considered planning and regulation.

Some of the ways that unimaginative or inattentive urban design
harms us is invisible. Stand on a street corner in any major city, close
your eyes, and listen. Brakes screeching. Truck cargo clanking. Engines
gunning. Fire engines and ambulance and police sirens piercing their
way into your ears. On any given day, in any given American city—
Chicago, Dallas, Miami, New York, Philadelphia, Phoenix, and San
Francisco—the ambient noise levels on busier streets significantly exceed the 55–60-decibel (dB) level of normal conversation, which is what
public health authorities at the US Environmental Protection Agency
and the WHO deem safe for everyday living. Noise levels on New York
City's subway platforms frequently approach 110 dB, which approxi-
mates the experience of standing three feet from a running power saw.

Harmful noise levels on New York City subway platforms

And while most public health authorities agree that urban noise levels should *never* exceed the jet-engine-taking-off-140 dB for adults (120 dB for children), they do. The European Union is no better: 40 percent of EU residents—whose countries offer among the highest general standard of living in the world—live subjected to noise levels loud enough to imperil their health and well-being. In developing countries, where noise protection ordinances are fewer and inconsistently enforced, the situation is immeasurably worse.

Hearing loss constitutes only the most obviously harmful effect of excessive noise. The World Health Organization outlines the detrimental effects of noise on other aspects of human health. As we've seen, it diminishes people's sense of control over their environments. Noise levels higher than 30–35 dB in residential neighborhoods disrupt people's circadian rhythms during sleep (even if they never actually wake up); disrupted sleep, in turn, contributes to a wide range of physical and emotional problems. Exposure to continuous environmental noise higher than 55 dB alters people's respiratory rhythms, and detrimentally affects our cardiovascular systems at noise levels of 65 dB or higher. When environmental noise levels exceed 80 dB (more or less equivalent to the sound of heavy truck traffic on a highway), aggressive behavior and vulnerability to mental illness increase.

Children who attend schools situated near airports consistently demonstrate impairment on a host of cognitive faculties that critically facilitate learning, such as concentration, persistence, motivation, attention to detail. With diminished capacity for reading comprehension, these children fall behind on achievement tests. Even an occasionally passing train disrupts children's ability to learn. One study compared the academic performance of two sets of students at an urban school located on a site adjacent to elevated train tracks. Those students whose

Stress, noise, and crowding in developing countries, Dhaka, Bangladesh

classrooms faced the train tracks consistently underperformed on a wide range of tasks relative to their peers in quieter classrooms located just across the hall.

Suburban Living, Suburban Landscapes

Suburbs purportedly offer quieter, more pastoral havens from the ills and travails of urban life: that's why people move to them. But the design of many suburbs impoverishes people's lives in other ways. Like Plano or Needham, many suburban landscapes discourage people from developing a meaningful sense of place. Why? For generations, most suburban developments have been and continue to be constructed by large regional or national real estate concerns: in 1949, close to 70 percent of the houses constructed in the United States were built by only 10 percent of the country's builders—and the home-building industry is more consolidated today than it was fifty years ago. Companies such as D.R. Horton, NVR, and PulteGroup pay minimal attention to a locale's climate, a site's topography, and to where they source their materials. The logic driving the site plans, building designs, and landscaping schemes they generate is profit, so their playbook is written to convention and maximum iterability in design, and in ease and speed of construction. Day in and day out.

Outdated land-use ordinances and building codes exacerbate these problems. Many municipalities retain zoning laws written in and for a different era, which violate best-practices urban design for the twenty-first century, such as community-oriented, sustainable, and mainly denser, mixed-use developments. Zoning laws that discourage density and cordon residential areas off from workplaces, light industry, and even retail establishments force even enlightened developers to overcome extra hurdles to make vibrant suburban communities. Anachronistic building codes discourage healthy experimentation, the widespread adoption of improved materials and newer, superior methods of construction. Again and again, innovation is forsaken for the ease of conventional compliance.

People who move to the suburbs in the hopes of a healthier lifestyle and more connection to nature can find themselves suffering problems they didn't anticipate. Suburban life remains primarily about the car: to get your kids to school, you drive; to get groceries, you drive; to get to work, you drive; to run errands, you drive. The typical suburban neighborhood, scaled to the view at thirty or forty-five miles per hour, and to the turning radius of the steering wheel, is designed for driving over walking or biking. One result is that many such developments promote lifestyles so sedentary and auto-dependent that America, and increasingly much of the developed world, is facing an avoidable public health crisis. As public health authority Richard J. Jackson bluntly puts it, "the more time we spend in a car, the more likely we are to be obese"— and this is without even taking into account the enervating, resource- and time-draining costs of commuting. An auto-bound, sedentary lifestyle is part of why nearly 40 percent of adult Americans are obese, and fully 70 percent are overweight—compromising people's cardiovascular health and general muscular capacity, and greatly increasing their vulnerability to type 2 diabetes.

In the name of privacy, suburbs can cultivate social isolation and promote the kind of social and ethnic homogeneity that breeds intolerance. Suburbanites, instead of rubbing shoulders with people of wide-ranging backgrounds, outlooks, and sensibilities, are deprived of the (well-documented) humanizing and socializing effects of being active in a diverse public sphere. They also lose out on the well-established psychic and social benefits of being enmeshed in closer and looser networks of people. And these failings are evident in suburbs across the country, from those that surround New York on Long Island and in Westchester County and northern New Jersey, to Dade and other

Highway commuting

"Boxed, bleached sameness": Las Vegas suburbs, aerial view

counties in Florida, to the vast sprawls that are Dallas and Phoenix, to Orange County and many other counties that make up the California megalopolises known as Los Angeles and the Bay Area.

One principal reason that people move to suburbs is well known: they wish to claim a little piece of nature as their own. Yet paradoxically, suburban developments do more to dissociate people from the varied experiences that a good natural environment provides. Psychologists have demonstrated again and again that people are invigorated and soothed by nature's compositions, but nature tamed by suburbia can seem like assembly-line "softscapes": simplistic, repetitive arrangements of shrubbery and grass draped over swaths of land feel more stultifying than uplifting. Theo, the protagonist in Donna Tartt's *The Goldfinch,* hilariously recounts his gobsmacked impressions of the exurbs of Las Vegas, where he unexpectedly lands: "I looked up and saw that the strip malls had given way to an endless-seeming grid of small stucco homes. Despite the air of boxed, bleached sameness—row on row, like a cemetery's headstones—some of the houses were painted in festive pastels (mint green, rancho pink, milky desert blue) . . . As a game, I was trying to pick out what made the houses different from each other: an arched doorway here, a swimming pool or a palm tree there." Later, Theo observes that "there were no landmarks, and it was impossible to say where we were going or in which direction." In the interior spaces of these cookie-cutter domiciles, the same sort of tedium exists on a smaller scale. The simulated woods found in the kitchen cabinets and flooring of such developments manifest little of the visual, textural, and olfactory complexity that people yearn for. Such features, all typical in suburban developments (and in many urban developments as well), constitute a catalogue of larger and smaller missed opportunities, rendering suburbs as inimical to human well-being as the urban experience they were supposed to rectify.

Neither cities nor suburbs consistently integrate considerations

Landscape design in most suburban developments barely exists

about landscape into their design. The public spaces outside buildings—lawns, squares, plazas, parks, cultivated and uncultivated outdoor areas wholly or predominantly made up of the greenery of trees, plants, flowers, and grasses—are much more often than not an afterthought, deemed unworthy of attention or design. In cities, we find perhaps a sculpture or a few sad benches; in suburbs, unvariegated lawns, punctuated by the occasional lonely bank of shrubs.

Boston's massive Central Artery/Tunnel Project—the Big Dig, as it became known—betrays the pervasive public apathy and governmental ineptitude when it comes to the design stewardship of our landscapes. The Big Dig dismantled the elevated part of Interstate 93 that cut through downtown Boston, putting the highway underground and restoring a strip of land that could have been used to knit the riven city back together. During its construction from 1990 to 2007, it was the largest, most expensive urban project in the United States, and its final price tag, more than $24 billion, made it the most expensive stretch of roadway in American history. Yet for years after its completion, the city of Boston and state

More planted than designed: Rose Fitzgerald Kennedy Greenway, Boston

of Massachusetts spent more servicing the debt for the project—$100 million—than was devoted in total to the entirety of the surface landscape that the Big Dig created. By commission or omission, nearly everyone involved in the project treated urban design, architecture, and landscape as an afterthought. Years passed before anyone even figured

out what to put there. As a result, close to a decade later, the Rose Fitzgerald Kennedy Greenway's public spaces, broken up by cross street after
cross street, are less *designed* than *planted*, making its formal name more
a term of self-parody than a descriptor. Six years after it opened, the *Boston Globe* reported that "one-third of the 20-acre Greenway is unfinished,
with some parks lacking basic furniture and signs, and parcels that were
slated to host museums and cultural institutions remaining barren"; a
recent visit confirmed that not much has changed.

Builders and Architects: Decision-Makers and Decisions

Boring buildings and sorry places are nearly everywhere we turn.
So, who exactly is in charge of our built environments? Who are the
decision-makers, and what kinds of information do they have available
to them? What kinds of decisions have been made in past generations?
What kinds of institutional structures are in place to which today's
decision-makers remain bound? And what kinds of actions could they
take that negatively or positively impact how we, our children, and our
children's children will live out our lives?

Among the groups exerting influence on the design of our built environments, the largest are construction companies, product manufac-
turers, and real estate developers. These are businesspeople, driven by
profits. Collectively, the construction trades constitute one of the least
efficient and most deplorably wasteful industries in the United States.
And because most construction companies, unlike companies in most
sectors of the economy, invest little in research and development, they
are highly averse to innovation. What's more, construction companies
and the manufacturers of building materials, fixtures, and finishes give
high priority to factors—for example, cost and ease of transport and
storage—which relate tangentially or not at all to the needs of a project's
eventual users. Construction companies and manufacturers of prod-

ucts for the built environment take an approach to design analogous to that of the highly profitable home-furnishings giant Ikea, which mandates that designers' products not only be easily assembled by consumers but also be easily stored in their colossal warehouses, with every product's component parts fitting into packages that lie flat.

Real estate developers, whether residential, commercial, or mixed use, contend with multiple constraints beyond labor costs, zoning, and building codes. What and how developers build depends upon the state of the economy and its projected growth, financing vagaries, municipal ordinances, the quality of available labor, the nature of the building products on the market. The time horizon on which developers operate is set by marketplace exigencies, and bears little to no intrinsic relationship to promoting quality design and careful construction. Some developers, like Jonathan Rose in New York City, may be driven by broader social goals—choosing impoverished areas to erect their projects, constructing sustainable, well-designed affordable housing—but the fact remains that real estate development is business. Without enlightened regulation and demand, the public good can be served in a sustained way only to the extent that doing so reaps a profit.

In coming chapters we discuss examples in many sectors, from manufacturing to retail to office space, of profitable, innovative, good design. Yet the current structure of real estate development mostly discourages high-quality products and experimentation. For nearly any project, securing financing, pulling permits, and executing construction takes months to years. The high interest rates that developers pay investors, usually banks, to finance a new project creates enormous pressure on them to complete a project as quickly as possible. All this perpetuates powerful incentives for developers to employ established site plans and ready-made building designs; to rely on familiar, readily available, off-the-shelf materials; to use them in the most conventional ways; and to settle for standard (and more often than not, actually substandard) construction practices.

What about designers? Surely they must be more attuned than developers to people's experiential needs. After all, most design schools train their students not only to minister to commissioning clients but to envision their role as guardians of the public realm, working to enhance the overall vitality of a city or place. Still, the reality remains that designers work for *clients* (which includes developers) and are subject to the same market forces as they are. The results of these market structures and limitations, and of the decisions of people working within them, constitute our built environments, not just in small-change residential and commercial developments but even in the highest profile projects. Consider, for example, what happened in the case of One World Trade Center, built on the hallowed Ground Zero in New York City's lower Manhattan. Designed by Skidmore, Owings & Merrill, architects of some of the most innovative skyscrapers in the world, including the Burj Khalifa and the Cayan Tower, both in Dubai, the many clients of One WTC (the Port Authority of New York and New Jersey, the developers Larry Silverstein and the Durst Organization) whittled away at the architects' design until by the time One WTC opened, it had devolved from a soaring, inspiring monument into three disaggregated components—base, shaft, and spire—that looked as though they'd been cobbled together by toddlers on a preschool floor. The glass shaft crashes into its clunky bunker of a base, which is scantily clad with glass fins like a prison clad in a sequined leotard. To make matters worse, at the eleventh hour, the spire was denuded by the Durst Organization of the elegant encasement the architects had designed for it. Too taxing on the company's owners' pocketbooks.

One WTC, like the London 2010 Serpentine Pavilion, shows that engaging the services of talented, highly trained professionals is no guarantee of success. Sometimes market pressures, transmitted through

The architect's rendering of One WTC tower (at left, Skidmore, Owings & Merrill), depicts a light base and an angled shaft, neither of which was executed as designed.

ignorant clients, are to blame. Just as often, though, designers simply lack sufficient knowledge about human environmental experience. This deprives them of their most compelling argument to advance good design—its powerful effects on human health and well-being—a lacunae that stems in part from the anachronisms of architectural education. Very few schools of design—with rare exceptions—offer, much less require, training in how humans experience the built environment. Courses in sociology, environmental and ecological psychology, human perception and cognition appear infrequently, if at all, in their curricula. The national accreditation boards do not mandate such studies as an essential part of professional training. Students learn about many basic and esoteric systems of design logic—complex geometries, structural systems, manufacturing and construction processes, parametric design. These are all important subjects, but only when combined with a curriculum that teaches students how the people who will use their projects will actually perceive, live in, and understand them. Of the built environment's effects on human cognitive and social development, students learn next to nothing.

Instead, design studios, where students receive direct training in how to conceive and realize a project, are often organized in a way that maximizes the competition for their professors' regard. The result is that students gravitate toward and are usually rewarded for dramatic, gestural, attention-grabbing forms, designs that pack the maximal visual punch. Unusual forms and statement architecture reinforce budding designers' impression that their projects are singular, discrete objects isolated from their environmental and urban context, and not intimately tethered to how actual people will experience and use them. Jeff Speck, an influential urbanist, explains the deleterious impact of such dynamics when it comes to conceptualizing very large projects: "Architecture students are taught that when you have a block-scale project, it's not only your right but your architectural obligation to make it look like one building. But walking past 600 feet of anything is less enticing than walking past a series of 25-foot segments of different things." A pervasive reliance on computer-aided design can further exacerbate the student's inexperience with scale that is a result of designing buildings as isolated, discrete objects. Drawing by hand—one means by which students can absorb the art of designing for the human scale and perceptual faculties—has largely fallen by the wayside.

When students enter practice, the emphasis on striking visual compositions that they learned in school is often perpetuated. In taking on a commission, whether it be an art museum or a sewage treatment plant, the architect, landscape architect, or urban designer faces a mind-boggling array of disparate tasks, many of them requiring different skill sets. She must learn enough to grasp the nature of that particular institution, listen to and interpret her clients' stated needs, and reality-test them against budgetary and other constraints. She must develop a comprehensive analysis of the site and its topography. She must acquaint herself with local building and zoning codes, construction practices, and materials. Out of it all, she must imagine a holistic solution, and then work out, step by step and beam by beam, how people can construct her design so that it becomes a concrete object in the world. Designing anything that will become an enduring feature of the built environment is a demanding, complex—indeed, a daunting—challenge.

The designer's literal task at hand is to create a three-dimensional, physical object or setting; her job—object-making—and the dictates of the design process train her focus on composing forms. But the design process is conducted mostly on a scale that is minuscule in comparison to the end product. In these circumstances, considerations of user experience—how a place works at the scale of the city or its site, how it comes to feel and serve its users over time and in different seasons, the granular details people notice when moving through spaces, the nonconscious responses people will have to a project's small-scale and less visible features, such as sound, materials, textures, and construction details—all this can remain in the lowlight.

The nitty-gritty of the design process inclines professionals to privilege a project's overall compositional and pictorial aspects, which is a tiny slice of how it will function in people's lives. Compounding the problem today is that professionals rely heavily on two-dimensional imagery—photographs, digital simulations—to market their services. They self-publish photographic monographs and construct elaborate websites to advertise their work. One highly respected contemporary architect, describing his process to me, confessed only half-jokingly that he always keeps in mind what his project will look like in photographs: "It's all about the money shot. That's what we design for." Why? Because that's what potential public and private clients, as well as jury commission members and colleagues will see and likely remember in making these decisions.

What information does a photograph actually offer about a building or space? Not much. Photographs confer the impression of veracity, but distort a multitude of compositional and experiential features. Pictures notoriously distort color, so that a concrete building you expected to gleam white could turn out to be a dull, thudding gray. Pictures fail to capture the ethos of a place: how it sounds or smells, or how its materials feel. Money shots appeal only to our visual senses, and only depict a single moment in the best light of day. But buildings and landscapes exist in three dimensions and are experienced in four, looking and feeling different at dawn than they do at dusk, different on gray days than on sunny days. How grossly photographs misrepresent built environmental experience, especially scale, was brought home for me when a Venezuelan-born acquaintance returned from seeing Lúcio Costa's and Oscar Niemeyer's famous capital complex in Brasilia. These are some of the most iconic monuments of late modern architecture in the world. In real life, she reported, those buildings looked as though they had

Photographs distort and prettify: Secretariat and Chamber of Deputies (Oscar Niemeyer), Brasilia, Brazil.

been built at half scale. "They were no bigger," she exclaimed, "than my health club in Caracas!"

Many, if not most published (and posted) photographs of contemporary projects are commissioned by their designers. These not surprisingly elide all sorts of information—indeed, anything that might hinder their solicitation of new clients. Pictures of Zaha Hadid's enormous, blazing white Dongdaemun Design Plaza, set in a green landscaped park, make it seem visually arresting, like some breathing, pulsing giant whale beached in a sunken plaza in downtown Seoul. Looking at the real building rather than photographs reveals that Hadid's complex is bedeviled by cracking concrete, misaligned joints, and a landscaping scheme so uninspired that what was once green has dried into an insipid brown. Similarly, photographs of another widely celebrated project, the Seattle Central Library by Joshua Prince-Ramus and Rem Koolhaas's OMA, do not reveal the building's self-absorbed hermeticism, its utter failure to connect to the street-level spaces, to knit it into the city's larger public realm. They mask its haphazard detailing. And they obscure the fact that the library suffers a dearth of quiet, comfortable spaces to read. The main reading room, located on the tenth floor, frustrates many over-forty readers hoping to settle in for a long day's work, as the nearest restroom is located three floors down. These examples could be multiplied by dozens, illustrating how photographs of the built environment, and the photographers behind the lens, present the illusion of verisimilitude while unwittingly and willfully misrepresenting precisely what they purport to depict.

A Self-Perpetuating Cycle of Sorry Landscapes

The cognitive revolution's complete rethinking of human experience reveals much that should make us all reconsider what we know about design. We respond to our environments not only visually but with our many sensory faculties—hearing, and smelling, and especially touching, and more—working in concert with one another. These surroundings affect us much more viscerally and profoundly than we could possibly be aware of, because most of our cognitions, including those about where we are, happen outside our conscious awareness. The leaders of the cognitive revolution reveal—perhaps mostly inadvertently—that our built environments are the instruments on which this orchestra of our senses plays its music. Just as we imagine waving with the conductor as we listen and watch the opening of Beethoven's Fifth Symphony, when we navigate and inhabit our environments, what and how we consciously *think* is inexorably bound up with our nonconscious simulations and cognitions, and with what and how we feel. Most surprisingly, our emotions, our imagined bodily actions, and especially the memories that we develop of them are embedded in our very experiences of built environments, and loom large in how we form our identities.

Most people don't think about the built environment much at all, and certainly not in any systematic way. Our buildings and city-scapes and landscapes command a vanishingly small platform in the media and the public sphere. How many substantive articles about the built environment does your favorite newspaper or website feature? And how many of those contain actual analysis rather than featuring simple presentations?

For many reasons, most people mostly ignore cityscapes and build-ings and landscapes. The two obvious reasons for this are that build-ings and streets and plazas and parks rarely impinge on our conscious experiences. They change slowly, if at all. And we are animals: neuro-logically, wired to ignore all that is static, unchanging, nonthreatening, and seemingly omnipresent.

We also mostly ignore our built landscapes because practically, we have no obvious stake or influence in their production. This aligns with our approach to other swaths of our daily experience and needs: for medical help, we go to doctors; to repair our car, we visit the auto me-chanic. Most of us, implicitly or explicitly, have relinquished control over our built environments, having entrusted decision-making about them to the putative or established experts: the city council members, the real estate developers, the builders and contractors, the product manufac-turers, and the designers. Most of us perceive ourselves as helpless to make changes in the built environment. This very sense of powerless-ness results in a paradoxical situation: real estate developers configure new projects based on what they believe consumers want, which they assess mainly by examining what previous consumers have purchased. But when it comes to the built environment, consumers gravitate toward conventional designs without thinking very much about them. So developers continue to build what they think people want. No one steps back to consider what might serve people better, what people *could* like, or what they actually might need.

Not only are consumers disposed to prefer familiar, conventional designs, they will prefer conventional designs *even if those designs serve them very poorly*—which, as we have seen, they often do. This is owing to a common psychological dynamic, namely that the more times a person is exposed to a stimulus, even if it does not serve her well, the more she will habituate to it such that she eventually will not only prefer it when offered other options, but will eventually deem it to be *normative*. In this way, people can and do come to judge inferior places that serve them poorly, or even harm them in covert ways, as indisputably, objectively *good*.

This self-perpetuating cycle of built environmental inertia in which we are caught must end. Thanks to the cognitive revolution, the news is in: impoverished cityscapes and buildings and landscapes impoverish people's lives. Do we really want the inertia of convention, the profit margins of lenders and developers, anachronistic zoning and building codes, criteria such as the width of roads and flatbed trucks, to determine how our cities and institutions look and function and feel? Shouldn't we be rethinking design of all kinds, including its standardization, in light of all we know about what people actually need?

In the wake of the cognitive revolution, we must recognize the reality that aesthetic experience, including our aesthetic experience of the built environment, concerns more than pleasure, so much more that the conventional distinction between *architecture* as the province of the elite, and *building* as the province of the masses, must once and for all be eradicated. From our perspective—the perspective of how human beings experience spaces, of how built environments affect our well-being—such a distinction is incomprehensible and pernicious. The more we learn about how people actually experience the environments in which they live their lives, the more obvious it becomes that a well-designed built environment falls not on a continuum stretching from high art to vernacular building, but on a very different sort of continuum: somewhere between a crucial need and a basic human right.

CHAPTER 2

Blindsight

Experiencing the Built Environment

Rationalists, wearing square hats,
Think, in square rooms,
Looking at the floor,
Looking at the ceiling.
They confine themselves
To right-angled triangles.
If they tried rhomboids,
Cones, waving lines, ellipses—
As, for example, the ellipse of the half-moon—
Rationalists would wear sombreros.

    —WALLACE STEVENS, "Six Significant Landscapes"

There are things
We live among 'and to see them
Is to know ourselves.'

    —GEORGE OPPEN, "Of Being Numerous"

A single potent metaphor encapsulates much about the built environment's complex role in our experience and our internal world. Blindsight is a disorder that robs a person of her conscious sense of sight: someone suffering from this condition will tell you and herself that she is blind—wholly, unremittingly blind. But it's not true. If you ask a blindsighter seated in a room where she thinks the light source might be, she is quite likely to point to the correct place. Ask her where an object in that room is located, and the number of times that her guesses are correct will substantially exceed chance. The neurological explanation for blindsight is that lesions in portions of the visual cortex prevent blindsighters from developing a *conscious* awareness that they can see. But other portions of their brains (the parietal cortex, superior colliculus, and thalamus) continue to register such information in a meaningful way.

For understanding how people experience the built environment, the case of one blindsighted woman is especially salient. She suffered a variant of this condition called left hemisphere neglect, and insisted that she saw nothing on the right side of her visual field. A researcher showed her two pictures of a house, in which the left half of both pictures was identical, but the right half of one—shown to her "blind" side and not visible to her "seeing" side—depicted the house engulfed in flames. Asked what she saw, the woman repeatedly maintained that the two images were identical. Yet when pressed to choose one of the houses as her home, she consistently selected the intact, nonburning house, even though she never could explain a reason for her choice.

When it comes to how we perceive the built environment, every one of us, even a professional, is more or less blindsighted: mostly oblivious to how our brains process the places we inhabit, all but completely unaware of how we integrate that information into our experiences, and largely clueless about how it orients our movements, affects our cognitions, our emotions, and our choices. Yet there's a difference. Blindsighted patients are neurologically incapable of being coaxed into becoming conscious of the visual information that their brains register. The rest of us are much luckier: we can learn to see.

The story of our relationship to our built surroundings is revelatory: rich, multilayered, and, owing to the changing rhythms of the day and the operations of human memory, temporally complex. Experiencing the built environment involves not just how we process the swirl of sensory cues and impressions at the moment that we apprehend them. It also involves the way that we subsequently store our cognitions, whether it be the conversation you had with the friend you ran into at a local café or the business deal you made over an office lunch. At such times, indeed at most times, what we think and experience seems wholly independent from the particularity of the place. But when we remember such events, we unfailingly access something about the environments in which they took place. So we need to understand some fundamentals about the complex architecture of cognition—how people initially process sensory and mental impressions, as well as how we remember and recall them—and that those fundamentals reveal that the built environment thoroughly permeates and is manifestly at the core of human experience.

The Architecture of Cognition: A New Paradigm for Understanding Built Environmental Experience

A new account of human cognition is emerging from the combined fruits of many research disciplines. At its core is knowledge derived from two powerful new scientific disciplines, cognitive neuroscience and cognitive neuropsychology, both of which were born only recently, in the wake of a spate of technological innovations that allow us to study the human brain and its functions with unprecedented insight and precision. Knowledge from these sciences is cross-pollinating with research in myriad fields, including environmental, social, and ecological psychology; artificial intelligence; behavioral economics; cognitive linguistics; and neuroaesthetics.

This still-evolving account of cognition already has begun to fundamentally transform the common understanding of human experience—that unified impression we take away from those moments of what we see, hear, and smell, as well as what we think, touch, feel, and do. Experience is grounded in our sensory perceptions and in our internal thoughts, which together govern how we make sense of the information that comes to us from being in the world. And when something happens in the world or in our minds, that "something" is always *situated*, in our bodies, in a given time, and in place.

One thing this new account of cognition reveals is how extensively our environments structure and provide the framework for not only *what* but also *how* we think. It turns out that we respond to our constructed worlds in some surprising blindsight-ish sorts of ways. Take these examples. If someone sitting inside a five-foot cube of a box tries to solve a problem, his solutions are likely to be less creative than when he tries to solve the same problem while sitting outside that box. Or this: if your partner is working on a spatial, verbal, or mathematical problem, you'd help her more by switching on the task light on her desk—bright idea, bright lightbulb—than by flipping on the overhead light. Or this: a real estate broker wanting to make a quick sale would do well to show her clients a living room with curved rather than straight-edged surfaces, because people respond to curving surfaces with "approach" behavior and in general tend to prefer them.

Who would imagine that where a person's body is situated relative to a large container would prompt him to think, or not think, "outside the box"? Or that creativity can be abetted by task lights more than by overhead ones? What these and other studies confirm—there are hundreds of them—is that *cognitions* constitute the core of any experience. Because we live at the same time in the world and in our heads, human experience contains the realm of the mind: it's our cognitions—which include our emotions—about events that constitute our lived experience.

And that, already, is a helpful way to reframe our subject, because while "experience" might seem inexpressibly idiosyncratic and amorphous, "cognition" is not. About cognition—the many processes by which people understand, interpret, and organize sensory, social, and internally generated data for their own use—we have much more than the intuitions, hypotheses, and hunches that have mostly guided us in the past. About cognition, we have knowledge.

We need to shift our thinking in three ways to properly explore the nature of cognition and its role in built environmental experience. First:

What our minds think is largely shaped and profoundly influenced by the human body that we have. Second: This, along with the fact that our human bodies are shaped by the environments in which we live and have evolved, suggests that much of our internal cognitive lives takes place outside of language and below the level of our conscious awareness. Third: These factors transform our understanding of how humans live in the world by making us less the imperially sovereign agents over our experiences that we often believe ourselves to be, and more the environmentally embedded beings that a bird's-eye view of any human settlement would suggest.

The new paradigm of human cognition begins by reframing the relationship of our thoughts to our bodies. Cognitions do not emerge in tension with a corporeal self, as was thought for centuries, nor from a disembodied mind—a paradigm encapsulated in the dualistic "mind-body problem." Instead, cognition is the product of a three-way collaboration of *mind, body,* and *environment.* Inherent in the very fact of human embodiment—life lived in a body—rests the notion that the physical environments that a body inhabits greatly influence human cognitions. The body is not merely some passive receptacle for sensations from the environment, which the mind then interprets in a somewhat orderly fashion. Instead, our minds and bodies—actively, constantly, and at many levels—engage in active *and interactive,* conscious *and nonconscious* processing of our internal and external environments.

All this is quite new. The common western understanding of human thought and experience relies on the idea, first formulated by René Descartes in the seventeenth century, that our conscious mind operates at least on some level independent of its corporeal home. The basic structure of this Cartesian dualism is as follows. First, through our senses—sight, touch, taste and so on—we receive information from the environment. After we sense a stimulus, we perceive it. After perceiving, we begin to process, forming a preliminary judgment about that information by running it through our internal data bank of familiar, recognizable patterns and by reacting to it emotionally. Thus we conjure a preliminary interpretation of the initial stimulus. Only then comes the highest step of cognitive processing, whereby we consciously use logic, reason, and abstraction to evaluate the importance of the given stimulus to our life and make decisions about whether and how to act. This still-dominant but soon-to-fade model of human cognition resides deeper in Western culture and people's minds than does any knowledge of Descartes; it serves, indeed, as a sort of folk model of cognition. It has led, and continues to lead, everyone—including ordinary people, makers of public policy, and designers of the environment, cities, and buildings—astray.

The emerging mind-body-environment paradigm starts differently: with the somewhat obvious fact that the human brain inhabits a body, and that this brain-mind-body lives on the earth, in space, and in the social world. The brain and the body together facilitate the operations of the human mind, which depends on their architecture for its very existence and for its modes of functioning. Human cognition takes place in a corporeal body that lives on the earth and in space. Not only that: our cognitions are shaped by the fact of our embodiment, sometimes in surprising ways—such as thinking more creatively when we sit outside (instead of inside) a box.

In this new paradigm, a cognition can be linguistic or it can be prelinguistic; it can occur anywhere on the spectrum from the nonconscious to the conscious. Learning to understand cognition's complex, multilayered, often subterranean quality involves attending to our own fleeting thoughts and perceptions—precisely the ones that we are more or less predisposed to ignore. And learning to understand human cognition in the built environment involves the same process, with an added focus on our surroundings.

Imagine going for a walk in a park. You are walking alone, lost in

Thinking inside the box

concentrated thought about an upcoming meeting. Now compare that
to your experience—same place, time, and route—when strolling with
a friend, lazily sharing an occasional observation about what you see
around you. Finally, compare these experiences to one in which you
are by yourself in a relaxed state, looking at blooming azaleas, inhaling
the park and its air, listening to singing birds or squealing children.
Regardless of what you are focusing on, thoughts and perceptions about
the built environment are with you. They are fleeting, easily drowned
out by other kinds of more insistent, louder cognitions, usually about
ourselves and our lives: our colleagues' support of a political candidate
we dislike, the projected fortunes of our favorite sports team, the flowers we smell, the noisy children we hear.

Those cognitions that are more audible, more distinct, usually come
in the form of the words we hear inside our heads. Language is the enabler and medium we use to express our internal thoughts to ourselves
as well as the enabler of social communication. Because words have
such a hold on us, many philosophers of language and thought have for

generations mistaken our interior monologues or the spoken language that forms them for the entirety of cognition. It is true that many of the cognitions *of which we are consciously aware* are indeed linguistically framed, including that ongoing monologue inside our heads. But now we know that many, many more of our cognitions are not verbal. And sometimes they even precede the words we conjure up to describe them.

Nonlinguistic cognitions include sensory impressions—cold feet, a breezy room, a knobbily textured rug—and fleetingly perceived patterns—a geometric figure, voids playing off solids. They also include the full battery of emotions and feelings—the comfort of a gently curving wall that embraces us. And they include patterns of associations, called schemas, which we mentally construct through our experience of growing and living in our bodies in the world. One example of a schema was revealed in the study showing that we associate the flash of a suddenly illuminated lightbulb with the internal experience of stumbling upon a new insight. A patterned set of associations—a schema—links the visual experience of seeing a light flick on with the more abstract notion of a flash of insight.

Cognitive scientists of all stripes, including contemporary linguists, now maintain that a good deal of what people think to make sense of our innumerable, fleeting sensory experiences, which we then use to understand more complex or abstract concepts, is not only *not* verbal but prelinguistic. As we grow from dependent children into more or less autonomous adults, ever in pursuit of goals in the face of daily challenges, we increasingly master our own bodies and develop ever greater competence in the world. In the process we collect an immense storehouse of mental schemas, which Mark Johnson, a cognitively informed philosopher, describes as patterns of "organism-environment interactions."

It is this storehouse upon which we rely, rapidly and without conscious effort, to navigate, interpret, and make sense of our physical environments and the objects they contain. Nonverbal cognitions, including schemas, come to us nonconsciously, at least at first, transpiring beneath that ongoing verbal monologue inside our heads. Standing inside Daniel Libeskind's extension to the Denver Art Museum, next to its ominously canted walls, provokes an immediate physiological response somewhere between unease and fear. Following this will be a nonconscious cognition—move away from that wall! The term *nonconscious* as we use it here, then, does not mean "not verbalizable"; it simply means "not in words, thought or spoken." Every ordinary person experiences nonconscious cognitions; with knowledge and focused attention,

Denver Art Museum (Daniel Libeskind)

we can bring our nonconscious cognitions into the kind of conscious awareness that allows us to articulate them verbally.

One of the cognitive revolution's most astonishing revelations is the extraordinary prevalence, even dominance, of nonconscious cognitions in people's lives: some estimate that as many as 90 percent of our cognitions are nonconscious! This means that most of us greatly overestimate the degree of sovereignty we have over our thoughts and the reach of our agency. Our conscious cognitions seduce us into believing that when we feel, perceive, and think something, we do so consciously, if not exactly deliberately. This distorted sense of our own control appears to be an essential tool enabling humans to prosper, because if all or even most of our cognitions were conscious, then complex and indeed even simple tasks would overwhelm us. As it is, the limited space of conscious cognition is reserved for important tasks. The result? We remain, in Daniel Kahneman's words, "blind to our blindness," enjoying an "almost unlimited ability to ignore our ignorance." And yet, all the while, swirling outside the domain of language, our nonconscious cognitions—including those about the built environment—are amorphous and omnipresent. A structured polyphony of half thoughts shapes our emotions, decisions, and actions every day.

Nonconscious Cognition in the Built Environment: A Walk in the Village

Imagine, for a moment, that you live in the West Village of Manhattan. Having awakened early, you follow your morning routine, readying yourself to leave for work. You open the refrigerator, notice that there's no milk for your morning coffee, and in an instant, decide to walk to the corner store. During most of the ensuing fifteen minutes, from

Streetscape, West Village, New York City

that "Shoot! No milk" moment to your handing over of money to the grocer's cashier, your conscious thoughts are engaged. A contentious telephone conversation you had with your brother the night before. The reservations you need to book for an upcoming trip to Chicago. Whether to urge your teenager to accompany you and your best friend on a weekend outing.

During those same fifteen minutes, myriad nonconscious cognitions are scuttering about in your mental world. Many concern your combined perceptions about and actions in your immediate surroundings—in other words, the built environment. The moment you realize that you've run out of milk, a fleeting mental image of the breezy, crammed interior of the neighborhood market, even perhaps of your arm reaching into the glass-enclosed refrigerator where milk is stored, wafts by. That's one kind of nonconscious cognition: mentally, you *simulate* yourself undertaking a rote-patterned action in the world by using impressions drawn from memory. Then you glimpse your apartment's front doorway. You know that, in order to be on your way to the store, you need only pass through that nearly seven-foot-tall, person-accommodating aperture that materializes when the door swings open. The mere sight of the door activates in your mind another nonconscious simulation of a familiar action pattern, departure. That simulation, like most, is *cross-modal* and *sensorimotor*, in that it simultaneously involves both your sensory faculties (here, vision and proprioception) and your motor system (moving your muscles in a coordinated fashion to get your body out the door). In this case your sensorimotor simulations have you heading toward the door, taking your jacket off the hook, retrieving your keys from the dish on the dresser, and extending your hand until it reaches the cool, smooth brass doorknob to turn it and pull the door open. That simulated action sequence, so routine that you barely need to think consciously about it, is also a schema. Such schemas, innumerable and ever-present, so pervade our built environmental experience that it is as though they are literally embedded into the structures and objects of the world, serving as cues to activate nonconscious cognitions.

The glimpse of the path to your front door, the eye-level shelf in the neighborhood grocer's dairy fridge, the door with its brass knob facilitating your way out of your apartment; we think of our sensory faculties as separate modalities, but most nonconscious cognitions draw from and are the combined result of impressions from a number of different sensory faculties: we can call them *intersensory* (involving more than one sensory faculty working in collaboration). Nonconscious cognitions about the built environment incorporate visual impressions in combination with impressions from other sensory faculties. Some of these are familiar—touch, hearing, and smell—and others less so, having been conceptualized in the last few decades without having broken into the ranks of the canonical five senses that children are taught (at least in the United States).

These less familiar though no less important senses include *interoception*, by which you monitor your sense of your internal body and the relationships among its parts. *Thermoception* relates to the discernment of temperature and the sensory response to it, imagined or real. The architect Alvar Aalto, building in his northern, native Finland, painted the staircase floors bright yellow and encased the handrails of his metal banisters in wooden sleeves, because he correctly intuited that people need only *look* at a wooden handrail in a sunny yellow stairwell to feel warmer. Perception is intersensory. *Proprioception* gauges your sense of your body and its parts in space, and helps you monitor the location of your body relative to the objects and spaces around you; it is the difference between visual and proprioceptive perception that creates the aesthetic power of the famous Palazzo Spada staircase in Rome, in which Francesco Borromini, the Italian architect, used forced perspective to make us anticipate that our ascent will be longer and more arduous than it actually is.

Haptic impressions are visual stimuli that provoke us to mentally

Opposite: Wood-handled banister in a bright yellow stairwell, Paimio Sanatorium (Alvar Aalto), Paimio, Finland

Stairs with forced perspective, Palazzo Spada (Francesco Borromini), Rome, Italy

Just looking at an excessively rough texture can provoke thoughts of retreat: Yale Art and Architecture Building (Paul Rudolph), New Haven

simulate tactile sensations: as with Aalto's banisters, the mere sight of them cues our imagined sensorimotor engagement with them. That's why cushioned chairs in a so-called soft classroom elicit feelings of relaxation and warmth even if students never sit in them. An example of how important haptic sensations can be to our overall impressions of buildings and places is Paul Rudolph's Yale Art and Architecture Building in New Haven, with its rugged, cobblestone-filled exterior and interior concrete walls. Many people dislike the A&A building, in all likelihood because they nonconsciously imagine that brushing up against it might hurt.

During your morning excursion for milk, you draw on the wide range of your nonconscious cognitions, the faculties that produce them, and the cognitions' palpable, intersensory immediacy in your internal world. Coat donned and door locked, you rely upon your interoceptive and proprioceptive faculties to navigate the familiar fluorescent-lit corridor leading from your front door to the entrance of your building. Stepping outdoors, you find your hand recoiling from the cast-iron banister you would otherwise use to steady your way down the icy stairs onto the sidewalk. On the street, your skin shivers from the bite of crisp winter air that is funneled through the neighborhood warehouses and townhouses. Along your journey down the street to the corner, you use your proprioceptive, visual, and sensorimotor faculties to skirt the steps of townhouse stoops, while your olfactory faculties make you nonconsciously recoil from the stench of smelly garbage, piled outside a nearby restaurant. You bypass, without really seeing, New York's "alternate side of the street" parking signs, hear without really hearing the drone of passing cars on the nearby avenue. Without deliberation or effort, your various sensory faculties stand on call, ready to collaborate as needed to help you navigate the uneven terrain of pitted sidewalks, fire hydrants, and scrawny, fenced-in trees reaching desperately toward the sky.

Nonconscious and conscious cognitions differ in strength rather than in kind. Stronger cognitions muscle their way into attentive recognition. Nonconscious ones, less demanding, nevertheless pulse on and are accessible to our conscious awareness, expressible in words, but only if we train our attention upon them. Philip Glass has likened composing music to slowing down long enough to listen to a subterranean river of sound that always flows in his head, whether he attends to it or not. Like Glass's generative river, nonconscious cognitions flow steadily beneath the level of consciousness, a river running wide with memories of and information about our own body states, the spaces of and objects in our environments, the patterned, schematized ways that we might interact with them.

During the whole process of deciding upon and taking your morning walk to get milk, the subterranean river of your nonconscious cognitions flowed, with you navigating by acceding to its flow. Standing in your kitchen before leaving, you glimpsed your apartment's front door in your peripheral vision. Without your conscious awareness, that glimpse turned your yearning for milk into a decision to venture out rather than settle for black coffee. If the front door hadn't landed in your peripheral vision as you stood at the open refrigerator, you may never have considered the journey or, had you considered it, you might have deemed a milk run too time consuming an undertaking.

Even when we pay no conscious attention to the built environment or focus only on selected aspects of it—that's nearly all the time—it functions, in our lived experience, as a never-ending concatenation of what some social psychologists call *primes*. A prime is a nonconsciously perceived environmental stimulus that can influence a person's subsequent thoughts, feelings, and responses by activating memories, emotions, and other kinds of cognitive associations. The sight of the front door sufficed to prompt your memory of traversing its threshold en route to exiting the building. The door, in other words, functioned as a prime, activating your imagined simulation of departure.

As you deliberate whether or not to go, you mentally simulate the route from your home to the corner grocery. This in turn permeates your now-conscious deliberations, and the simulation itself influences your impulse toward exit by making it seem less effortful. You know the route well, and people consistently estimate familiar journeys to be shorter than unfamiliar ones of equal distance, probably because the latter require more mental exertion to navigate. You decide to run out.

Consciously, what you will remember most about the morning is that you spent it mulling over the previous night's fraught telephone conversation with your brother, and your conclusion that the hurtful things he said were unjustified. That's why you experience your walk to the store as having been unpleasant. You may or may not recall the frigid wind tunnel rushing around the surrounding buildings, or the revolting stench of uncollected garbage. But what's very unlikely is that you would connect your harsh judgment of your brother's conduct to the physical discomfort you felt as you were considering his sour remarks.

Body position influences mood: forcing your lips into a smile creates a "smiling" feeling

That people experience emotions first as physical states—as *feelings*, in other words, as things that we feel in our bodies—and only then as cognitions has been hypothesized ever since one of the founders of modern psychology, William James, proposed it. We now know, for example, that the cerebellum, which coordinates sensory input with muscular responses, is also involved in processing emotions. Fear manifests itself as a jolt of energy, and muscles tense. Disappointment is embodied in slumping shoulders. Smiles beget happiness, and happiness smiles. It goes both ways: if you assume a body and facial position associated with a certain emotion—look up while forcing your lips into a smile, for example—you are likely to feel "smiley" inside. Today, psychological research confirms that what we call "feelings" are cognitive responses to what our bodies literally *feel*, and not just in the case of the familiar fight-or-flight response activated by the feeling of fear. Our emotions are enmeshed in and intermeshed with our bodies; in other words, they are "in the body," or *embodied*.

Often we register this connection nonconsciously, which can lead us to erroneously conflate an embodied emotional state with a cognition that objectively is unrelated. Here's how it works. Your block-long walk from your apartment to the neighborhood grocer subjected you to stinky garbage and a tunnel of frigid winds. You blanched; you held your breath. You experienced enough physical discomfort from the cold that your muscles further tensed, as you closed in on yourself in a protective stance. In all, you assumed a physical demeanor associated with feelings of contained anger and a distressing sense of isolation. By chance, your bodily clenching up—a reaction to physical discomfort—coincided with the moments in which you were consciously musing upon your contentious telephone conversation with your brother. And that coincidence of environmental, bodily, and cognitive events had an important consequence: you judged your brother's words more harshly than you might have in other environmental circumstances. Just as being in a bad mood about work can make you more impatient with your partner, so too does bodily position—which the built environment influences—affect your moods and even your actions.

In this way, our cognitive interpretations of emotional states can profoundly influence our conscious state of mind. Had the weather been warmer or the buildings on your street designed so that they did not produce a wind tunnel; had the garbage bins been hidden out of sight instead of just off the sidewalk in an alley, you might well have considered your brother's comments with less pique and more compassion. Nonconsciously, you causally related your felt body state of physical discomfort with an emotional state—anger, distress—that happens to be physically manifested in the same way, and this in turn influenced your cognitions.

Another way to say this is that wind tunnels and garbage heaps functioned as primes. Our built environment is riddled with primes, and because it is, a design can be deliberately composed to nudge people to choose one action over another. Some of the primes by design you encountered during your walk in the West Village include visual axes, resulting from the placement of walls and other functional elements in your apartment (between the refrigerator and the front door); spatial sequences, or corridors of movement (the straight shot from your apartment to the grocer—having to turn corners would likely have changed your calculation, regardless of the actual distance between them); the massing and composition of buildings (which, for example, resulted in a wind tunnel). A change in a visual axis, a spatial sequence, in the way that solids are massed and volumes composed could prime very different cognitions.

Primes and Spatial Navigation: Orthogonal and Hexagonal Grids

In the worlds that we inhabit, every building element, every sequence of voids, every surface, every construction detail could potentially prime our cognitions. Yet, of course, not everything we encounter becomes a prime—and at any given moment, most elements in our environments have no influence upon us at all. How, then, does the nonconscious mind select or settle on a given environmental feature or become suggestible to its influence? There are myriad answers to this crucial question. As a preliminary step, we can say this: our subterranean, nonconscious cognitions about the built environment rarely respond predictably or well to design features and elements that fail to take our human embodiment into account. Such an assertion may seem so obvious that it hardly bears mentioning, and yet, the global array of sorry places and boring buildings suggests otherwise. Countless elements, objects, and features in and of our built environments—homes and schools and offices and parks and roads—have been designed and constructed with little attention to how they accord with the architecture of human experience.

The rectangular or square grid is a good example. Pragmatism largely explains why the grid so extensively pervades the history of design. Before digital computation, designing with straight lines and right angles greatly reduced the complexity of construction and facilitated engineering and systems of construction because simple, regular measurements determined the placement of structural supports. The arrangement of rooms and paths and corridors, of solids and voids, could all then follow from the grid's transparent logic. In the late nineteenth and twentieth

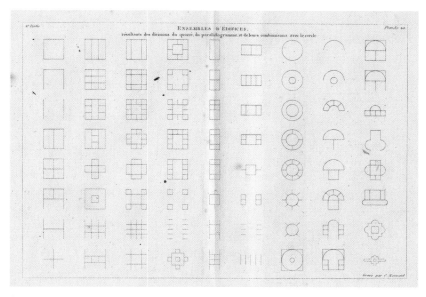

Grids, Jean-Nicolas-Louis Durand, *Partie graphique* (1821)

centuries, when the manufacture of construction materials came to be dominated by mass production, the grid became especially useful; to this day, commercial buildings and especially contemporary skyscrapers often employ a five-foot gridded module, from which the dimensions of floor plates are calculated and interior spaces designed. In large-scale residential housing projects, the modular dimensions differ, but the grid nonetheless remains a widely used template for design.

Architects have championed the practicality of the grid since the early 1900s, when the influential French pedagogue Jean-Nicolas-Louis Durand developed a system and taught generations of students that a building of nearly any size and complexity could and should ideally be designed along a modular square grid. Early modernist architects such as Walter Gropius, entranced with the possibilities of mass production, reinterpreted Durand's design for the fabrication and construction of model projects for affordable housing, in the Weissenhofsiedlung in Stuttgart, Germany, which opened in 1927. Gropius employed the grid for the floor plans, interiors, and exterior facades of his Houses 16 and 17, arguing that this method would bring down the housing's overall price tag by lowering the costs of manufacturing and transporting building materials, and by simplifying its construction so greatly that it could eventually be built by unskilled labor. Contemporaries of Gropius also championed the grid but for different reasons: Ludwig Hilberseimer advocated its use not only in small single-family homes but also in tall residential apartment complexes and in urban plans, maintaining that the resulting homogenization of building design and construction would help modern urban dwellers—who tend to be nomadic, moving from place to place—feel comfortable wherever they went.

Gropius and Hilberseimer were certainly correct from the point of view of construction and design: the ease and practicality of the grid is incontestable. (Just look at the plan of any midwestern city, or of Manhattan.) But throughout the history of modern and contemporary architecture, the

Above: Grids in houses: Weissenhofsiedlung, House 16 (Walter Gropius), Stuttgart, Germany

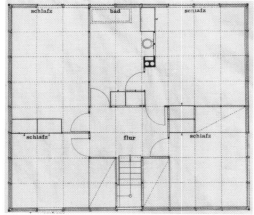

Above and Left: Grids facilitate construction: Weissenhofsiedlung, House 17 (Walter Gropius), Stuttgart.

Below: Gridded Cities: Ludwig Hilberseimer, from *The New City*

grinding simplicity of buildings and cities composed with perpendicular straight lines intersecting at right angles has been condemned and even caricatured. One commentator on Baron Georges-Eugène Haussmann's replanning of Paris under Napoleon III, a construction project that involved cutting massive straight boulevards through the city's medieval fabric, derisively joked that Haussmann would "arrange the stars above in two straight lines" if he could, and nearly a hundred years later, the Italian avant-garde group Superstudio published a series of imaginary land and cityscapes illustrating the grid's dehumanizing effects.

Today, studies on the cognitive techniques that people use to navigate their way through spaces have elucidated why some designers have felt a persistent sense of unease when faced with the grid's pervasiveness. Spatial navigation is a complex process. In order to get us safely from one place to another, our brains rely on the collaboration of place recognition cells and grid cells in the hippocampus and parahippocampal region; these help us to continuously update our position vis-à-vis the objects around us, a system poetically referred to as "dead

Grids do and don't fit people: *Images of Life* (Superstudio)

reckoning." But the grids that our brains construct in dead reckoning are not right angled. Instead, to navigate our bodies through space, our brains nonconsciously imagine a *hexagonal* lattice of points, and locate the place of our body with reference to two objects in space, forming an equilateral triangle within the hexagonal grid. From any single given point, neighboring fields will be located at sixty-degree increments.

Armed with this knowledge, compare Gropius's Weissenhofsiedlung Houses to Frank Lloyd Wright's Hanna House in Stanford, California. Wright also was concerned with the problem of creating well-designed, affordable homes. But he opposed Gropius's determination to modify the automobile manufacturing industry's mass production techniques to build low-cost housing. Building hundreds of homes for people of limited means, Wright eschewed the simple rectilinear grid. In the

Wright saw "things out of the corner of his eye": Hanna House (Frank Lloyd Wright), interior

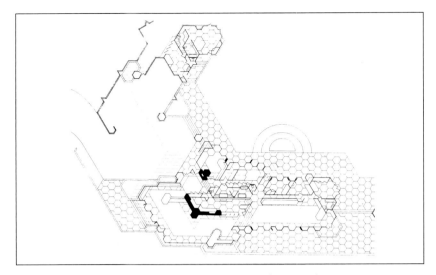

Triangulating the location of our bodies with two points (or objects) in space, our brains use grids made not from squares or rectangles, but from triangles and hexagons, to help us navigate through space: Hanna House, plan

home he designed for Paul and Jean Hanna in 1936 in Stanford (now owned by Stanford University), he employed a rather unconventional geometry of equilateral triangles arranged into a hexagonal field.

These shapes echo those in natural forms such as honeycombs and soap bubbles, and because they do, Wright thought, people would find them intrinsically—in other words, nonconsciously—appealing. Perhaps. But Wright also intuited that people would be drawn to spaces arranged according to hexagonal geometries because they are consonant with the dictates of human visual perception and might well facilitate a more effortless spatial experience. "When Dad builds," Wright's son once said, "he sees things out of the corner of his eye." Cognitive neuroscientists Edvard and May-Britt Moser and John O'Keefe have now confirmed what Wright intuited: human spatial navigation is organized around our practice of nonconsciously, imaginatively triangulating the location of our body in space with two other proximate points, in order to help us find our way.

Right-angled grids have their virtues and will always have an important place in the built environment. But pragmatism no longer dictates
their incessant, unmitigated use. Designers now are able to execute designs that are not just mass-produced but also mass-customized around
the exigencies of human experience, because of recent developments
in the technology of computer-aided design and computer-aided manufacturing. A project's overall composition and component parts can be
more complex and more specifically tailored to the site, the users, and
the functions housed (called a project's *program*) than what had been
technically feasible in the past.

Two Ways of Apprehending Environments: Direct Responses and Metaphorical Schemas

The nonconscious cognitions that continuously flow through our minds
as we navigate our built environment come in two kinds of responses,
direct and indirect. Direct responses are physiological, not learned. In
this case, a given feature in the environment *in and of itself* provokes in
us a rapid, automatic response, such as when we recoiled from the smell
of garbage, felt a heightened sense of agitation in Nouvel's Serpentine
Pavilion, or detected a subtle sense of fear in Libeskind's museum. The
most obvious direct responses are orchestrated by the brain's amygdala,
the seat of fear, which produces our so-called fight, flight, or freeze response.

Anyone who has visited a haunted house or mounted a "Doom
Buggy" to journey through one of Disney's Haunted Mansions has experienced the powerful automaticity of the human direct response system. No matter how sober we are, no matter how insistently we remind
ourselves that it's all staged, nothing but a performance, logic proves a
weak ally when we find ourselves trapped in a dark, shadowy place with
no discernible boundary or measure, or in a room with no apparent exit,

or standing atop an unstable floor that may give way and precipitously drop us into space, or hearing sudden, unexpected movements or loud, unexpected high-pitched noises. Objectively, visitors in the Haunted Mansion know that such stimuli pose no danger. But that never stops us—indeed, it *cannot* stop us from feeling viscerally, palpably afraid, even if the dissonance between the physiological fear that we feel in our bodies and the conscious, cognitive knowledge that we are not imperiled results more in thrills than terror.

Other examples of direct responses include our vague sense of discomfort when we find ourselves in environments proffering too little or too much information. Overstimulating environments exhaust us: the Chancery Court clerk's suffocatingly cluttered office in *Bleak House*; a crowded underground transit passageway that ignites our urge to scan constantly. These are compulsions that likely emerged during the tens of thousands of years when humans lived outdoors and needed to keep vigilant watch for any movement that might suggest danger. Understimulating environments too—such as the suburban development Theo describes in *The Goldfinch*, or a poorly designed school's internal corridors—enervate us, so killing us with boredom that they exacerbate stress, sadness, and even addiction. We want to escape, to flee to a more cognitively engaging and healthful place.

Not all of our direct responses to places are triggered by fear or other negative emotions. Some places elicit direct responses that we experience with pleasure: paths promising unknown places excite our curiosity, and curved planes impart a sense of ease. Perhaps the most widely studied cluster of direct responses involves the effects of color on our emotions. People so consistently find both cool and less highly saturated colors calming, and both warm ones and more saturated colors energizing, that if they take tests administered in rooms that prominently feature red elements, they score less well. Color has all sorts of unexpected effects on us. When people took IQ tests in a room painted with a sky-

Retail interior in red

Retail interior (Prada) in pale green

blue ceiling, their scores were higher. Most famously, a certain hue of pink is reputed to be so calming that some football teams paint it on the locker room walls of their opponents. Much about how we perceive color varies across peoples and cultures, to be sure. But some basic color perceptions, like these, consistently activate direct responses, which vary least because they are deeply rooted in human biology. Maurice Merleau-Ponty, a philosopher whose writings anticipated the new paradigm of cognition, articulated this complex reality when he implicitly chastised proponents of radical cultural relativism in *Phenomenology of Perception*, advising that it was "time to stop wondering" how and why red, the color of blood, "signifies effort or violence" and green, nature's chosen color for plants, signifies "restfulness and peace." Instead, Merleau-Ponty continued, we should devote ourselves to rediscovering "how to live these colors as our body does"—in the actual environments that we inhabit, day in and day out.

Some of our nonconscious perceptions are indirect: they originate not in human physiology but in the cognitive schemas we construct throughout our lives as we learn to inhabit the world. When a glimpse of the front door sparked your thoughts of exiting the premises, when you mentally imagined reaching into the dairy case with your hand to grasp a carton of milk, when you estimated the duration of your projected journey to the grocery by calling up a cognitive map of your neighborhood, you drew on schemas that you'd created from past experiences. If you were staying in the same apartment but it was located in a city unfamiliar to you, the very same exit door might not activate the consequent visual, motor, and navigational schemas.

One large category of such schemas generates cascades of associative and often nonlogical cognitions. You certainly know that concrete and steel are static, heavy, and hard, just as surely as you know that water ripples, bubbles, and flows. The pergola at New York City's Central Park Lasker Pool, the bubbly roof-walls of PTW's National Aquatics

Lasker Pool, Central Park, New York City

Center in Beijing, dubbed the Water Cube, and the wavy profiles of Zaha Hadid's London Aquatics Centre are indisputably *constructed*: the roofs in New York and London are concrete structures, while a steel space frame supports the rigid plastic walls in Beijing. Even so, through design, these buildings stir up a cascade of associations that are logically unconnected to their physicality. The zigzagging horizontal line of the Lasker Pool's concrete pergola might elicit thoughts of water's rippling flow. You cannot help but think of bubbles if you find yourself looking at the Water Cube on Beijing's Olympic Green. And the sweeping lines of Hadid's Aquatics Centre evoke our sense of water's fluidity. Design conveys something of the experience of being in these places.

All these are instances of embodied metaphors in the built environment. Typically, people conceive of a metaphor as a poetic device. But

London Aquatics Centre (Zaha Hadid), London, England

Water Cube (National Aquatics Center, PTW Architects with Arup Engineering), Beijing, China

the word simply refers to a dynamic whereby we transport any kind of content or meaning—visual, sensory, auditory, linguistic, proprioceptive, interoceptive, or any combination thereof—from one place or thing to another. Etymologically, the word "metaphor" derives from two ancient Greek words, *meta*, which means to move over, across, or beyond, and *phoreo*, which means to carry over, and these roots indicate that metaphors are cognitive devices that people can use in many media and forms.

Let's say that you are looking for an apartment in an unfamiliar city. After finding one that you like, you call a friend to share the news. She asks why you like it, and you casually reply, "I feel *at home* there." That's a metaphor. You might not even recognize it as such. But you've taken the concept of a home, which denotatively is any large, constructed object containing habitable spaces in which people reside, to signify something else: the sense of emotional comfort and bodily ease that accompanies domestic inhabitation. How will your new residence in a strange neighborhood affect your sense of internal equanimity? In answering that question, you've *carried over* meaning drawn from one domain—the built environment—to another—your felt sense of well-being. We imbibe such metaphors out of our experiences of living in the kinds of bodies humans have, in environments both natural and constructed. That's why they are embodied.

Metaphors are schemas. They constitute a particular (and large) category of schemas whereby our experiences of familiar and concrete things are harvested to convey abstract notions, feelings, and ideas. Patterns go on wallpaper and rugs, yet we say, "She exhibits a disturbing *pattern* of distress." Rhythms originate in music, yet saying "He liked the *rhythm* of his days" or "The rhythm of windows in that skyscraper is unusual" would confuse no one. Tactile sensations feed into statements such as "That proved a *rough* ride."

Just as the rippling and flowing lines in the swimming facilities in

New York and London convey something of what it will feel like to experience those buildings, the "at home" metaphor conveys more about how you feel in your new apartment than would the simple statement that you "feel comfortable" there. Metaphors elicit emotionally salient associations, visual images, bodily sensations, auditory memories, and more. Everyone knows what it feels like (or is supposed to feel like) to be at home. Metaphors illuminate a wide range of abstract concepts—in this case, the feeling of safe well-being—by associating them with concrete, easily imagined images—in this case, the familiar "no place like home."

When they are effectively deployed, metaphors in the built environment can function as primes. The bubble motif at the Water Cube cues us to associate certain aspects of water, such as its buoyancy, ephemerality, and movement. At the same time, the motif glosses over and deemphasizes water's negative attributes such as wetness, cold, heaviness, and potential danger. This skewed fit between target and source, or in this case between the viewer's associations and the physical building or the functions it contains, is precisely what makes metaphors so effective. They accentuate critical aspects of a building's functions or experiences by highlighting and exaggerating them, while leaving us space for personal associations and interpretations.

Like the "at home" example, our most resonant metaphors originate in our early childhood experiences. As infants, we learn a vast array of abstract, even ephemeral ideas and concepts by associating them with familiar objects, patterns, experiences, and behaviors. A common metaphor conveys the importance of bigness. Throughout history, across cultures, people associate large size with physical and social power. George Lakoff, a cognitive linguist, and Mark Johnson dub this the *important is big* metaphor, which originates in our universally shared experience of infancy, when our caretakers were both much larger (a physical trait) and more powerful (a social and relational trait) then we were. They protected us. They played with us. They swung us up high; they scooped our tiny bodies up to gather us into their arms. Such metaphorical associations are the common property of all humanity, shared simply by virtue of our living in the particular kinds of bodies that we do. Other examples of metaphors that resonate with our experience in the built environment include *substantial is weighty*; *up is good, down is bad*; and *places* [the White House] *are events* [the political administration of a nation]. Wherever we live and whenever we have lived, people learn and have assimilated a large pool of basic, schematic metaphorical concepts, and they do so in the same developmental sequence.

Given how metaphors work, it is no surprise that they pervade the built environment. Take *important is big* and *substantial is weighty*. Throughout human history, governmental institutions, structures, and value systems have varied tremendously, in monarchies, totalitarian or theocratic regimes, communist or democratic countries. But a government, whatever its character or institutional structure, succeeds only if it endures, and part of the way that it does is by constantly reminding citizens of its substance, durability, and importance. As long as governments with resources have existed in human history, they've enforced the enduring power of their reign by constructing buildings that are tall, wide, and heavy. *Important is big; substantial is weighty*. So big and heavy is how governments build, from the pharaohs' pyramids in Egypt to Sultan Kunburu's Great Mosque of Djenné in Mali to Hadrian's Pantheon in Rome to the billion-dollar competition among nations today to build the tallest building in the city, country, continent, hemisphere, world. Thus, when we come upon exceptionally large buildings on a city walk, such as in Beijing's Tiananmen Square, which includes the massive Great Hall of the People, we immediately, nonconsciously, draw on schemas we internalized as children. *"Important is big." "Substantial is weighty."* Nobody misunderstands the social message conveyed.

People's association of physical scale and size with social power

Trenton Bath House (Louis Kahn and Anne Griswold Tyng), Trenton, New Jersey

informs how architects design even less august buildings. Through proportion and design, even the smallest of buildings can impart a sense of august gravity. Louis Kahn and Anne Griswold Tyng's bathhouse in Trenton, New Jersey, is nothing more than an open-air, protected room where boys and girls can wriggle into their bathing suits on their way to a summer swim. But Kahn and Tyng monumentalized this tiny building by exaggerating its weight and suppressing all exterior markers of scale. The Trenton Bath House features no base, no cornice line, no windows, no doors. Unbroken planes of undressed concrete block define geometrically pure, square, hollow prisms capped by low-slung floating pyramidal roofs. The important is big metaphor joins forces with related metaphorical schemas that are also based in grounded cognition: heavy is substantial; heavy is enduring; heavy is important.

People's metaphorical equation of the experience of physical weight with the perception of social power found confirmation in a recent psychological experiment. Some subjects were given heavy clipboards and others light ones. All were given the task of interviewing candidates for a new position. The subjects holding heavier clipboards judged their candidates (who were equally qualified) to be more intellectually and professionally substantial, and more appropriate for the job. Kahn and Tyng, manipulating proportion, drawing on metaphorical schemas associating bigness and weight with durability and importance, imbued this little concrete block building in Trenton with a dignity and timeless presence that belies its modesty and pedestrian function.

Intended and Unintended Experiences: Berlin's Holocaust Memorial

The Trenton Bath House and the Haunted Mansion show that the ways people apprehend the built environment—through direct responses and through the schemas of metaphors—can be manipulated or managed in ways that advance commercial, social, or indeed many kinds of agendas. But doing so takes skill, sensitivity, and knowledge: our cognitions and metaphorical associations are complex enough that orienting them by design can challenge even the most conceptually sophisticated architects. This is evident in the Memorial to the Murdered Jews of Europe, or the Holocaust Memorial, which sits on a 4.7-acre site near the Brandenburg Gate in downtown Berlin. Designed by the American architect Peter Eisenman, the Holocaust Memorial is composed of 2,711 concrete slabs arranged in an orthogonal grid, which vary in height but not in their eight-foot length or three-foot width. The slabs sit less than eight inches off the ground at the perimeter of the site, and increase in height as they progress toward its midpoint. At their tallest, the gray concrete slabs rise sixteen feet high.

Looking at the Holocaust Memorial in pictures or from surrounding city streets, and especially from an elevated vantage point conveys the impression of a terrifying monotony. At first its appearance in the

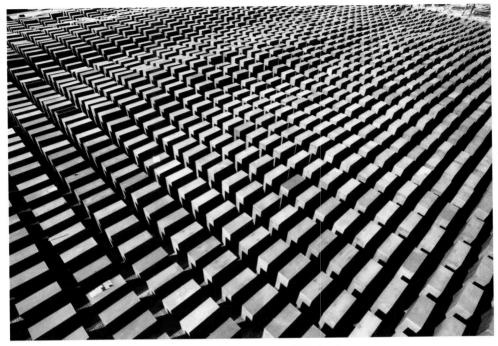

Aerial view of the Holocaust Memorial (Peter Eisenman), Berlin, Germany

People relaxing at the Holocaust Memorial, Berlin

heart of a vibrant metropolis, not far from the seat of German government, the Bundestag, looks like someone's disorienting mistake. What's this? Perhaps these relentlessly organized, carefully placed slabs are placeholders for some future construction project? Such perplexity was part of the designer's intention: he wanted to convey the sense of rationality—embodied in the monotony of the grid—gone mad. From this perspective, Eisenman was successful.

When we experience the Holocaust Memorial from the sidewalk and from within its repetitive slabs, however, our emotions and impressions are quite different. At its periphery, the memorial's low-slung, irregularly arranged concrete slabs pool into little eddies of space, where groups of visitors comfortably congregate. To step into the space of the memorial is to oblige an invitation gently offered. As we venture farther in, these gray monoliths rise in height while the paved ground plane slopes downward, with the memorial's lowest point near the center of the site. The designer meant this variation in the heights of the ground plane and the slabs to create the sense that we are descending into some utterly unknown place while feeling progressively more surrounded, trapped. But the moment-to-moment experience of walking through the Holocaust Memorial evokes less the horror of dehumanization than the thrill of the Haunted Mansion. Any discomfort we might feel is mitigated because its solids and voids are arranged along a rectilinear grid. Never do we feel the fear of utter disorientation, because the memorial's view corridors offer sight lines at predictable and frequent intervals, leading our eyes back to a city teeming with humanity—busy streets, crowded sidewalks, people on their way to life, people living their lives. Wherever we find ourselves in the Holocaust Memorial's ocean of dolmens, axes appear, graciously indicating exit routes, inviting us to explore yet another avenue. Never do we feel an ominous sense of threat, never trapped. Moreover, since the Holocaust Memorial has been (and was always expected to be) a popular tourist destination—people congregate there, bagged lunches set on the lowest slabs while children jump from one to another—we are never alone. Often, and sometimes almost literally, you find yourself running into other visitors, in strange if sociable encounters.

Eisenman conceived the Holocaust Memorial with embodied experience and embodied metaphors in mind. The restlessness of understimulation. The disembodied logic of the orthogonal grid. The uneasiness of nonvolitional descent. He intended his deliberately repetitious gray slabs to evoke something of what Jews experienced at the hands of the Germans during the Holocaust, when they were conceived of as numbers, not people, dehumanized and systematically murdered. But Eisenman's intentions do not translate into the built reality because his use of the grid as the datum by which to arrange the concrete slabs undermines the very experience it is meant to effectuate. His attempt to spatially manifest the metaphor "oppressively repetitive" lapses into its semantic opposite: "reassuringly patterned."

The Holocaust Memorial fails to effectively convey Eisenman's stated agenda for two reasons. He neglected to take into account all the different kinds of stimuli people would be processing when they actually experienced the site. Proprioceptively, they would sense themselves descending while the slabs loomed ever higher above their bodies, but visually, they would always use the view corridors to maintain their sense of spatial location. He also failed to recognize that in cities, buildings and landscapes and places are constituted not only by their formal composition but also by the social life they facilitate and structure: he designed it as if every person sees herself in isolation, never as part of a social group. As a result, Berlin has no memorial to the victims of a grisly, terrifying, history-altering genocide. What it has instead is an outsize folly that pretends to do the job, while children playfully jump its slabs, lovers steal kisses behind walls, and office workers picnic on the knee-high benches that await them. Visitors to the Holocaust Memorial, coming upon this urban playground, can't but note the incongruity of intent and outcome.

The Situated Essence of Autobiographical Memory: Places Become Us

Monuments like Eisenman's in Berlin are physical markers that people conceive and erect with the explicit goal of commemorating a significant historical personage, event, or institution. Their target is human memory. What, then, is the relationship of human memory to physical place? Are such monuments effective, and does their design really matter? Human memory—especially long-term memory—is one of many arenas of cognition in which our understanding has advanced dramatically in the past decades. Whereas scientists once thought that the brain contained some kind of discrete storage facility for long-term memories, we now know this to be false. When remembering a certain event from our past, we recall images, patterns, and impressions from many systems of sensory cognition, scattered in many parts of the brain. We also now know that these memories, which in sum help us to constitute our past, can be consolidated only by being linked up with our cognitions about physical locations and place. Put another way, what we know about how memories are consolidated in the brain reveals that the physical environment we inhabit during a given experience centrally figures in the memory itself. In the contemporary world, where our environments are overwhelmingly *built* environments, what this means is that the buildings, landscapes, and urban areas we inhabit are central to the constitution of our autobiographical memories, and therefore to our sense of identity. Our very sense of *who* we are and have been is inextricable from our sense of *where* we have been and are.

Recall a clear childhood memory, or your most triumphant adolescent moment, or your first day at your first adult job. Focusing on that memory, you will remember how you felt: your sense of well-being as you and your admired elder brother collaborated on the design and construction of a cardboard box fort, your elation when a teacher complimented your work in front of the class, your concentrated eagerness on that first job to prove yourself a meritorious hire. Then ask yourself if the chosen memory comes to mind blank? Very likely not. It probably came embedded in physical places and spaces: the people you were with, the sights you saw, the sounds you heard, the tactile sensations you experienced. Recalling an autobiographical memory *includes mentally simulating something of the place* where it originally occurred. This is why students perform better on final examinations administered in the same room in which they originally learned the material being tested.

Only recently have scientists been able to explain the relationship of memories to place. Neurologically, the kinds of long-term memories that are autobiographical are consolidated, or prepared for long-term storage, in an area of the brain called the hippocampus and the adjacent parahippocampal region. Working with other areas of the brain, these parts of the brain also facilitate our ability to navigate spaces. In forming such memories, the brain may use not just the same general region that helps us to identify places, but the very same cells—the place cells. Place cells enable us to both identify a place and consolidate a long-term memory. So important discussions you had with your mother about the person you've decided to marry or with your boss about a promotion are mnemonically encoded with information about where the news was delivered—at your parents' home, sitting on the front hall steps; on the boss's couch across from her desk. This place-bound nature of long-term memory might, incidentally, help to explain why people do not retain long-term memories before around age three: it is only then that our spatial navigational strategies mature.

Here, then, is another stunning fact about human cognition that crucially informs our understanding of how people experience the physical world: we cannot recall a memory from our past without revisiting at least some elements of the place where the original event occurred—if not consciously, then at least nonconsciously. What follows from this is that *place-bound experiences constitute the very framework for our sense of self and perceived identity.* The built environment constitutes the foundation upon which our past, present, and future selves are constructed.

This process by which we retrieve an autobiographical memory suggests also that the sensory components we associate with a given memory will powerfully influence the meaning we assign to the new environments we encounter. As we search for a new place to live, we might choose an apartment where we happened to see the sunlight hit the floor because it reactivates our memory, along with all its attendant sensations and impressions, of constructing that fort with our brother, with the late-afternoon light streaming through a window at a similar angle. That treasured recollection, along with the sense of intimacy we felt from collaborating with our beloved sibling, will influence our impressions of the new place. And every time we retrieve that memory, we strengthen the associations we draw between place and event, further inscribing it in our brain, connecting the existing memory to new internal and external stimuli. In this way our experiences, and then the memories of our experiences, are necessarily, fundamentally embedded in environments.

What does this new account of autobiographical memory mean? To say *the built environment is us* is but a slight exaggeration. And it is certainly no exaggeration to say that the built environment shapes who we are and how we move through the world physically, socially, and cognitively, as well as in the sense of how we construct and reconstruct our identity. An example from my own life demonstrates the point. For most of my first fourteen years I lived with my family in Princeton, New Jersey, on a sleepy street of architect-designed homes sensitively sited on wooded, generously landscaped one-acre lots. Martin L. Beck, a professor from Princeton University's School of Architecture, designed our

Autobiographical memories come packaged by place: 74 Allison Road, Princeton, New Jersey

house, and he had been influenced by Frank Lloyd Wright's vision of the modern home as a domain of quiet refuge nestled into the natural world. Our family residence, although small by today's ever-expanding suburban standards, offered a cherished retreat. Its dark wood-stained facade allowed only an occasional view onto the front yard and street, with an entrance recessed deep in shadows. But on the opposite side, the rear of the house, a panorama of green filled our view. Sliding floor-to-ceiling glass panels opened onto a large, entirely contained backyard from which no other home was visible.

Every member of my family understood the special pleasure of living in these idyllic surroundings, screened off from our neighbors by flowering dogwoods and magnolias, forsythias and flower beds, while being just a short, quiet walk from the university and Princeton's main shopping street. Inside, the house's wood-paneled interiors emitted a warm, orangey-brown glow, with spaces loosely arranged around a central fire-and-staircase core, offering quiet, private places to sit, read, or play. Even sitting alone in contemplative solitude, we always knew where the other family members were in the house.

This house—my parents sold it forty years ago!—continues to remain integral to the person that I am today. The sense of privilege we all felt, living in that Princeton house, shaping even today how I inhabit and think about my current neighborhood of East Harlem, New York. In that Princeton house I developed an acute sensitivity to the comparatively less fortunate circumstances in which many others—indeed, most people in the world—live. My appreciation for the powerful effects of a well-designed place—for architecture's capacity to afford security, ease, and well-being; for nature's soothing presence—originated not from a textbook but from personal experience. That house was alive, lush with nature: its public spaces filled with daylight and the landscape's ever-changing views, with shadows providing places in which a child could spirit away to hide. Just thinking of it suffuses my body with a warm, relaxed sense of well-being.

Whatever your particular childhood memory, you in your body and you in a specific place are the glue that binds together your sensory impressions, your thoughts, your emotions, and the meaning of all of it for you as a person, then and now. If you'd been raised in that Haitian shack or in the Needham McMansion or in a penthouse apartment on the moon, you'd be a different person from the one who is reading this book. The obvious and not-so-obvious implication of this is that the multistage process of experiencing and recalling the built environment underlies how we come to understand who we and who others are— namely, in place and in space.

This new understanding of autobiographical memory has profound and wide-ranging implications for how we understand human experience in general and human experience of the built environment in particular. If our understanding of others, our understanding of the world, and our understanding of our very selves are all inextricably enmeshed in our physical environments, then the importance we should accord the built environment and its design in our lives, societies, and politics becomes well-nigh unbounded. The cognitive mechanics of autobiographical memory *resituates the built environment inside us*: it constitutes the internal architecture of our lives. Far from being a backdrop or merely some marginally important stage setting that we can ignore without consequence, the worlds we build construct the literal, actual scaffolding we use to cognitively construct ourselves as people, other people as human beings, and our relations with one another. Inevitably and logically, what follows from this is revolutionary. To a nearly incalculable degree, the sorry state of our built environments compromises and impoverishes in pervasive and concrete ways our lives, the lives of others, and the lives of our communities.

This is why the places we build need to do much more than fulfill basic human physical, biological needs for the primitive function of shelter. They centrally figure into just about everything about who we are, who our children are and will become, and who we see others as and as becoming. Because the built environment is integrated into our self-identity and conceptions of others, it plays an *active* and *central* role not just in how we construct ourselves and our pasts, but also in how we, singly and together, move forward in the world. The design of the built environment is important for how we act in the present. It is important for how we will conduct ourselves in the future.

Thus far we've discussed the built environment as a totality. But built environments are wholes comprised of parts. Streets have curbs and sidewalks and stoops and streetlamps and paving. Buildings come with windows and roofs and thresholds and backs and fronts. Landscapes can be urban plazas, botanical gardens, water reclamation sites, or playgrounds with trees, playground equipment, and fountains. Understanding how these structure our experiences, our cognitions, and our identity necessitates examining them also at the microscale. After all, everything humans have built and will construct ultimately serves people. And the people who inhabit the edifices they construct live first of all in bodies, in bodies that stand on earth.

CHAPTER 3

The Bodily Basis of Cognition

I measure myself
Against a tall tree.
I find that I am much taller,
For I reach right up to the sun,
With my eye;
And I reach to the shore of the sea
With my ear.

—WALLACE STEVENS, "Six Significant Landscapes"

T hat mind and body are radically intermeshed can best be understood, perhaps, by those who have grappled with the problem of how to care for someone who is in the last years of life. Imagine an elderly relative, maybe your mother, who has lived for many years in a large, rambling suburban house where the closest supermarket and bank are a ten-minute drive away. She can no longer safely drive. She can no longer comfortably climb stairs. It has been nearly a decade since she's been able to properly manage the responsibility of maintaining property. Her deteriorating physical capacities indicate the prudence of your relocating her to either some assisted living facility or an in-law apartment near you.

Resist the temptation if you can. Elderly people tend to be better off staying in long-familiar habitats than moving to unfamiliar surroundings. Not only will your mother live more happily in her own home, but it's likely that her health will be more robust, even if the medical care she'd receive at a new place is equivalent to what she gets now. A more practically appointed, conveniently situated house would ease her daily burdens, yet your mother's emotional and physical health could deteriorate more rapidly. Why? Because her *cognitive* experience of leaving her longtime abode will deleteriously affect her body's *physical* health. Mind and body are one: what she thinks about relocating to new surroundings trumps the pragmatic benefits they'd offer.

Today all major scientific accounts of human cognition, whatever their differences, are predicated on the *integration*, not the separation, of mind and body. Our minds are profoundly enmeshed with our bodies, which structure the mind's thoughts in both form and content: our nonconscious cognitions, the schemas and the metaphorical associations we construct from living in our bodies over time; our memories; our emotions; and our conscious cognitions. Our cognitions are all bound up with the fact that we inhabit the very specific kind of human body that people have, and human cognition transpires only as *embodied* cognition. But what exactly do we mean when we say that cognition is embodied?

Embodied Minds

Look at any conventional representation of the human form. I've chosen one of my favorites, a fresco depicting Adam and Eve's expulsion from the Garden of Eden by the Italian Renaissance artist Masaccio. Here are bodies of the kind we've seen countless times before, including standing in our own mirrors. Top to bottom, whether we focus on Adam or on Eve, the head measures about an eighth of the total size of the body, which looks more or less symmetrical along a vertical axis from the crown of the head to the soles of the feet. Adam's and Eve's legs and torsoes claim the most area, and their hands, fingers, feet, and toes the least. Each eye measures no larger than a pillbox.

The design of the places we inhabit can and should comport with the Adam and Eve dimensions and characteristics of human embodiment. That is incontrovertible. As Masaccio poignantly depicts, naked, slouch-

How others see our bodies—
allocentrically: Masaccio, *Expulsion
from the Garden of Eden*, 1425

ing Adam and Eve need protection from their harsh, imposing world: clothing, shelter, or both. In the shelters they will inhabit, entrances must be of a height and width that will accommodate the maximum dimensions of an ambulating human.

But knowing that human bodies are vertically oriented and bear two arms, two legs, two eyes, and ten fingers gets us only so far if we want to investigate how we, cognitively, experience ourselves as we live embodied as actors, and embodied in geographically concrete places. The new accounts of how we think demonstrably elucidate that how we experience our bodies differs from how they exist as objects in the world. Put differently, the relationship of our thoughts to the bodies that we inhabit does not exhibit the isomorphism that is suggested by the bodies of Adam and Eve and what we see when we stand in front of our mirror. We do not experience our lives as we look in the mirror; rather, we look at the environs with our eyes, sense them with our skin, hear them with our ears, move around in them, and reflect upon it all as we do so and then again afterward. How we experience those bodies and minds of ours can diverge considerably from what that mirror presents to us. A simple example: in our experience of our bodies, our fingers, which measure small, figure much more prominently than our hips.

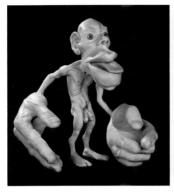 

Sensory homunculus   Motor homunculus

To envision how we internally and mostly nonconsciously experience our bodies—really, our body *schemas*—and how we in our bodies
experience the world, a different artistic representation of the human
body comes closer: the ridiculous-looking creatures, familiar to cognitive scientists, called homunculi, or "little men." A homunculus is
nothing more than a topographical representation of the brain's two
adjacent central stations for processing stimuli, in the motor cortex (the
right figure in the illustration) and the sensory cortex. Combined, the
sensory homunculus, which maps our sensory faculties, and the motor
homunculus, which represents our motor abilities, depict how we experience our bodies in the spaces and places of the world, mapping in
three dimensions the real estate that your brain allocates to processing
information from and about each of your body's various parts. More
technically, the homunculus is a representation of a person's entire
body surfaces and sensory faculties, draped on the postcentral gyrus in
the right cerebral hemisphere.

If we compare the head of the motor homunculus, for example, to
the head resting on our own two shoulders, we see immediately that
its proportions are fantastically distorted. The motor homunculus confronts us with protuberant eyes and nose, grotesquely huge hands, and a

mouth many times larger than his feet, all contrasting to a ridiculously, risibly scrawny torso and toothpick legs. (In the sensory homunculus, ears and genitalia make an appearance too.) This is how our minds experience the relative importance of the different parts of our body in space. So, within our conventionally proportioned Adam and Eve bodies, we mostly make our way through our days as cognitive homunculi, devoting far more attention to the information gathered by the homunculi's largest parts. They contain the greater concentration of nerve endings. Our eyes and ears deliver to our minds far more information about the world than our shins do. Our sensitivity to stimuli from the places that are big in the homunculus is so exquisite that comparatively, its small places are insensate dullards—that's why the paper cut that stings your finger would barely register on your thigh.

That we come embodied in Adam and Eve casings conveys one sort of information about our human experience: it describes our experience of other people, and the mental schemas we construct of our incarnated selves when seen or imagined from the *allocentric* (from outside ourselves) point of view. A fuller account necessitates that we consider the *egocentric* (from inside ourselves) point of view, the mental schemas we generate of our own body from our life of days standing, sitting, moving, and sleeping inside our bodies. These allocentric and egocentric schemas are not our mind's only embodied representations of ourselves to ourselves. Third, fourth, fifth, and sixth schemas are engendered from our experience of our bodies depending on where we direct our attentions, which skills we develop, and what we do at any given moment. A dancer's body schema differs from that of a taxi driver or a football player. My own body schema differs when I climb a flight of stairs with my middle-aged knees from when I listen contemplatively to Philip Glass's *Satyagraha*, a treasured piece of music. These differ from the schemas I employ when I'm threading a needle to sew a button on my daughter's sweater or working out a taxing math problem in my head. This third through nth body schemas are skill-based and task-oriented, specialized bodies that represent how we look to our minds when we are engaging in our varied activities.

That cognition is situated in the body, or *embodied*, has myriad implications for understanding how we experience the built environment. How our minds operate and what they register depends upon the anatomy of our human body and on the technical operations of our sensory and motor faculties: the colonies of nerve endings at our fingertips; their relative absence on our derrieres. The perching of the head atop the torso, the placement of the eyes in the head, the ways those eyes process line, contour, angle, light, shade, and color. All of this enables and constrains what we see, hear, feel, and think. For example, the human head, with its eyes and ears and nose and mouth, contains many of the faculties that critically enable us to interact with the world. Because of the fact that our heads are placed up high in our bodies, especially relative to smelly feet and the waste disposal system located below and behind us, our mind considers objectionable things to be "beneath" us. Here is another example: experientially, when we enter a workspace covered by a ceiling measuring eight feet high, its spaces will feel palpably, substantially more confining than when we enter an office with ceilings that rise up to ten feet. And yet we would barely register the difference between a twelve-foot-high ceiling and a thirteen-and-a-half-foot-high one, because when the height of a room's ceiling substantially exceeds our vertical reach (composed of the height of our body with an arm fully outstretched), our capacity to gauge its measure diminishes.

Bodies Permeate the World

Everywhere, our environments feature the traces of our allocentric bodies. When architects calculate the height of a door, the depth of a window seat, the proportions of a corridor, or the sight lines of an auditorium, they take into account how our human forms and sensory and motor faculties determine the ways in which we occupy and engage with our physical environments. But occasionally, designers incorporate the subtleties not only of our bodies but also our embodiment into a more comprehensive and deeper approach, with results so compelling that they alone make a case for more sustained attention to these nuances.

A poignant example comes from the traditional Japanese residence. The tatami mat measures approximately three by six feet, a dimension that makes the mats portable. More significant, it comfortably envelops most Japanese bodies in slumber. The tatami mat's short end, in turn, corresponds to the three-foot-wide module of the sliding wall and door panels in a traditional residence. These proportional relationships—slumbering body to mat, mat to wall, wall to standing

The dimensions of a traditional Japanese tatami mat fit the human body

Designed for many ages and modes of experiencing: Geopark (Helen & Hard), Stavanger, Norway

body—result in beautifully human-body-centered, iterative design systems.

These principles can be extended to apply to the body's scale and proportions at different stages of human development. When Helen & Hard, a Norwegian firm, created their sensational Geopark for a sweeping, disused waterfront site in Stavanger, the country's largest oil-producing and shipping hub, they designed for people's *egocentric* experience by taking measure of younger children's bodies for a bubble-ball-filled jumping area and birdcage-like communal swing, while shepherding adolescent bikers and skateboarders toward a larger, gently inclined terrain park. But they also designed for people's *allocentric* experience, mitigating the extreme difference in scale between the monumentally sweeping harbor and the comparatively diminutive human body by installing a playfully rhythmic arcade built out of recycled plastic mooring buoys, anchored atop tall metal stands.

The built world must be composed for the allocentric human bodies that use them; that seems obvious. The design of a chair should emerge from a careful analysis of how people position their bodies when they engage in particular activities while seated—for example, dining, reading, relaxing, dressing. And yet even this only begins to capture the complexity of good design, because chairs should do more than simply accommodate our human, Adam- and Eve- shaped bodies. They must adapt also to our egocentric embodiment. Marcel Breuer, a Hungarian-born American architect, designed one of the most famous, visually resolved pieces of furniture of the twentieth century, the Wassily Chair, by bending metal bicycle tubing into a continuous geometric shape that was light but still strong enough to support the human body. Despite the Wassily Chair's popularity (especially among architects), it not only looks mechanical but feels cold, as Aalto, Breuer's colleague, complained, arguing that it was designed for the allocentric body but not for the exquisitely sensitive, boggle-eyed and large-handed sensory homunculus. Aalto maintained that such chairs fell short of humane design because "a piece of everyday furniture in the home should not reflect too much light"; nor should

"Mechanics of Sitting: Dining, Reading, Relaxing, Dressing." Different activities require different body positions, *Aalto: Architecture and Furniture* (Museum of Modern Art)

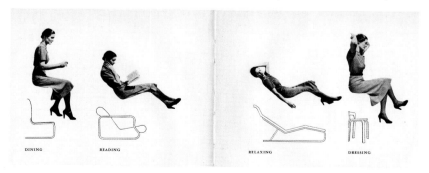

DINING   READING   RELAXING   DRESSING

Bent tubular metal looks and feels cold: Wassily Chair (Marcel Breuer), 1926

an object that comes into "close contact with the skin . . . be made of a material that is an effective heat conductor."

Aalto's counsel on furniture is ignored in equal measure when it comes to the design of our neighborhoods and urban spaces. A typical person, on her typically sized legs, will comfortably walk a quarter of a mile in five minutes, and many people will choose walking over driving if they are able to reach their destination in fifteen minutes or less. City planning theorists and sustainability advocates maintain that this information translates into a sensible algorithm: urban neighborhoods should be organized so that children can reach their schools and their parents can reach their workplaces and obtain all the household's basic goods on foot. Yet how many American and Asian cities violate this maxim! From Houston to Beijing, superblock dimensions and suburban dispersal maroons us in neighborhoods that fail to comport with the embodied humans that people are.

Harmonizing the physical environment with human embodiment

can take more subtle and evocative forms. Architects can play off our discrepant body schemas, allocentric and egocentric, in ways that are enlightening, engaging, and surprising. One common strategy is to exaggerate the difference between the slow, linear path that we pace out with our feet, step by step, and the fleeting, holistic representations that we create with scanning eyes. Japanese architect Tadao Ando, in his Poly Grand Theater in Shanghai, slashes through floors and ceilings to make interior spaces external. As we approach the building, Ando offers us glimpses of people occupying parts of the building that are unreachable from where we stand, foregrounding the divergence between the path of our feet and the paths of our eyes. Rem Koolhaas and the Office for Metropolitan Architecture concocted an even more surprising moment for visitors to the Kunsthal in Rotterdam. When we stand in a downstairs gallery, absorbed in looking at the art and therefore experiencing ourselves primarily egocentrically, we cannot but sense a

Showcasing the difference between the path of the feet and the eyes: Poly Grand Theater (Tadao Ando), Shanghai, China

rustling of movement above our heads. An upward glance toward the transparent ceiling reveals the soles of shoes and, behind them, the dramatically foreshortened bodies of art viewers standing in the gallery located directly above our heads. Suddenly, catching sight of what our bodies look like as objects viewed from below, we become aware of our embodied selves in relation to other people and to the object-filled, allocentric world around us.

To what extent do the features of our cognitive alter ego, the homunculus, govern how we experience the built environment? Remember that our eyes, ears, nose, hands, fingers, lips, tongue, and feet are the primary receptors for the sights, sounds, smells, and so on from the outside world, and that all these sensory faculties collaborate with our motor system to enable us to walk while maintaining our balance, to touch and interpret tactile information, and so on. The homunculus's features show that some parts of our bodies play much greater roles in our experience than our allocentric bodies suggest they might. Ears akin to satellite dishes suggest that even the subtlest auditory stimuli—the rustling of paper in a quiet room—will have us listening. Because of the soft, nerve-rich pillows of skin on our fingertips, even a second-hand experience of texture—the ominous look of razor wire coiled atop a beachside fence—will make us recoil, owing to haptic perception.

Such sensitivities should thoroughly inform, indeed permeate the design of our urban spaces, buildings, and landscapes, and occasionally they do. Among the notable contemporary practitioners attuned to both our allocentric and our egocentric body schemas, Peter Zumthor may be the most self-conscious. Zumthor, a Swiss architect, embraces multisensory experience as a critical component in his endeavor to create emotionally resonant places. This necessitates designing a project from its overall form to its subtlest, tiniest, and even occasionally invisible details. In his one-room St. Benedict Chapel in rural Switzerland, Zumthor considered the quality of a user's visual, auditory, and tactile experience as he made decisions about the wooden building's shape, structural articulation, and materials. In this hillside chapel, he specified that spruce planks be laid crosswise atop a wooden sub-floor because the resulting flexion allows the floorboards to give and creak when people walk on them, a nuance that confers the experience of being in a much older building. "Interiors," he explains, "are like

Designing for seeing, hearing, touching, and moving: St. Benedict Chapel (Peter Zumthor), Sumvitg, Switzerland

large instruments, collecting sound, amplifying it, transmitting it elsewhere. . . . Take a wonderful spruce floor like the top of a violin and lay it across wood. Or again: stick it to a concrete slab. Do you notice the difference in sound?"

A visit to the small, boat-shaped St. Benedict Chapel is unforgettable. We transverse an alpine landscape so ravishing that it practically commands us to concede that nature never did need us and never will. When we reach the hillside hamlet, we step off a dirt path to enter a sanctuary that subtly registers our presence. Our every footfall updates us as to the location of our body in this space. Visitors may never become consciously aware of such a detail. But our own heavy steps become the hushed sound track reassuring us that we are there, nowhere else, priming us with an awareness that, by our very presence, we transform this place.

Le Corbusier, the Swiss-French architect, orchestrates a different but equally affecting concerto of body and building at Notre-Dame du Haut, his extraordinary pilgrimage chapel in Ronchamp, France. Inside this dark, cave-like embankment in textured concrete, windows look as though they were hand-dug into primevally thick walls. Many of the colorfully glazed apertures seem almost habitable, as if they were inviting us to climb in, reassuring us that they, as fortified bastions, will reveal to us the beauty of the outside world while shielding us from its perils and tribulations.

These projects by Zumthor and Le Corbusier are exquisitely attuned to the idea of human embodiment. Most places people inhabit are not. Consider the proliferation of artificial surfaces, from the silk plants in shopping malls to the plastic "wood grain" veneer on retail countertops to the fiberglass "masonry" on building facades. All these impoverished design choices imitate the superficial appearance of the materials they refer to while lacking their density, their texture, or their smell, or failing to approximate the way they absorb ambient temperatures and re-

Designing for looking, imagining, and feeling: Notre-Dame du Haut (Le Corbusier), Ronchamp, France

spond to air currents or sound waves. Because our eyes register more than our noses can smell and our hands and fingers and feet will touch, designers misguidedly hope such dismal fakeries will fool us into feeling the positive emotions we associate with our long-term memories of moments when we engaged with the actual materials they costume. But our bodies know better. Such stratagems, because they misunderstand the complexity of human environmental experience, are doomed to fail, leaving our multisensory imaginations cold. As a result, we live with countless ersatz surfaces that fail to excite our homunculan engagement with them because we, nonconsciously if not consciously, intuit that we are being offered poor, flat, lifeless simulacra. Most of the time, richness and resonance comes with authentic textures and surfaces.

To more fully understand how our bodies shape what and how our minds think, we can look in greater depth at three environments that differ from one another in function, form, and intention. The designers and artists of a cityscape centered on a monumental public sculpture in Chicago, a museum of local history in Antwerp, and a Gothic cathedral in France manipulate human embodiment's principles and features to various and quite divergent ends. Not one had a deep understanding of the new science of embodied cognition, but in different ways, they all give credence and life to Aalto's observation, confirmed by Zumthor, that on occasion, human intuition can be astonishingly, "extremely rational."

Bodies and "The Bean": Affordances, Peripersonal, and Extrapersonal Space in Chicago

Installed in 2006, Anish Kapoor's celebrated 110-ton sculpture, popularly known as The Bean but actually (and aptly) entitled *Cloud Gate*, sits in Millennium Park between the shoreline of Lake Michigan and the skyline of downtown Chicago. A visitor turning off Michigan Avenue passes through the boxwood hedges into the rectangular courtyard where *Cloud Gate* sits. An enormous, abstract, curving, mirrored stainless steel sculpture confronts us. Our senses of proprioception (which helps us locate our body in space) and vision come alive, captivated by this reflective, elliptically regular blob measuring sixty-six by thirty-three feet. Before entering this courtyard, we'd noticed mainly the park's entrances, the pathways, and people: that's because the human cone of vision begins only slightly above eye level and stretches most of the way down to the ground. Now the immensity of this singular object in the landscape, its alluring reflectivity and softly curving surfaces, sends our eyes upward. We cannot help but pivot our heads toward the sky. And just as the mere act of stretching our lips into a smiling position—no matter our mood—will imbue us with the pacific emotions that elicit a genuine smile, cranking our heads to look up activates the internalized metaphorical schemas that we associate with

Cloud Gate ("The Bean," Anish Kapoor), Millennium Park, Chicago

assuming such a bodily stance. Even though *Cloud Gate* is small in comparison with the surrounding tall buildings, it effectuates the "important is big" metaphor so effectively that we confer upon this mute sculpture some kind of distinction whether we do so consciously or not. On the axis from which we approach it, *Cloud Gate*'s concave mirrored surfaces gather together and present far more of Chicago's cityscape than we could apprehend in a glance with our naked eye.

Cloud Gate uncannily imparts two contradictory impressions. The sculpture momentarily disempowers us—unbidden, it has taken us by surprise. Yet at the same time, *Cloud Gate* enhances our sense of agency, since it relies on our participation to create the spectacle of its perpetual remaking. Our small and large movements, our glances this way and that, all change what appears to us on *Cloud Gate*'s surfaces. From our simultaneous sense of powerlessness—we've been caught up short by this unusual object—and agency, our ability to change what and how we and other spectators witness, a productive tension results.

Most of the time when we walk around a city or urban area, we experience our body egocentrically. *Cloud Gate*'s lilting, lazy curves attract us—remember, people gravitate to and tend to imbue curved surfaces with positive meanings. It sustains our attention by creating puzzles for us to solve. Its gentle curves resemble a body's curves, so its highly machined surfaces activate thoughts of living things. Its surface reflects the image of our own bodies and the bodies of the people around us. And paradoxically, while mirrors normally serve as tools for the private moments of observing your Self as the body that others regard, *Cloud Gate* transforms that intimate act of self-regard into a public and social act. For all these reasons *Cloud Gate* inspires an ineffable yet powerful sense of transgressive if benign excitement. We are challenged, indeed forced, to encounter our body publicly and privately all at once; allocentrically, from without, and egocentrically, from within. *Cloud Gate* flashes a startling beam of light on our subterranean expectations about the shape of our body—which it transforms in the most theatrical way—and about our body's relationship to its surroundings—the city—which, normally, we experience as straight-edged, static, and mostly recessive. With every movement, with each step, our body, the bodies of other spectators, and Chicago's skyline transmogrify. We are guest to and host of a fun-house spectacle, the likes of which we've never seen.

Our visit to *Cloud Gate* casts many aspects of the body-mind-environment relationship in high relief. In its mirrored surfaces we see our own allocentric body—titillatingly distorted—as we know that others see it. The distortions make us aware of the dissemblance between our allocentric and our egocentric bodies: when the size of our head grossly inflates in comparison with our fantastically minuscule torso, we become aware of how our egocentric body schemas delegate to our body parts their relative importance. Over time, as we put together this succession of wildly distorted images of our and other people's bodies, we build a thought-provoking and amusing experience that is sure to take root as a long-term memory.

People often think of the body metaphorically, as the container for the Self, which extends into the notion that a person lives "inside" her body surrounded by the "outsides" of other people and objects. At *Cloud Gate*, we and other people become insides and outsides all at once. We, the people we see, and this machined, ever-changing sculpture are all gently curving objects in space—morphic equivalents, seemingly, pulsating with life. Most of the time, our Self experiences our body as a stable component of our identity. *Cloud Gate* unravels the assumed unity of body and Self, flashing, then shedding countless distorted, fleetingly comical iterations of ourselves and others.

One of *Cloud Gate*'s central subjects concerns the human body's relationship not so much to itself as to the surrounding landscape. Just as Masaccio's Adam and Eve differ from the homunculus, how we think our body looks changes relative to the external spaces in which we imagine or experience it. We know that the human body, seen allocentrically, simply constitutes one object among many objects—animal, vegetable, and mineral. The space that all those objects occupy can be mapped out using theoretically identifiable coordinates.

This, however, reveals little about how we experience the space of our body in the spaces and landscapes of the world. From the egocentric space of our body, the world more resembles a continuously changing montage of disparate stimuli. Lived spaces are three- and four-dimensional carnivals of opportunities exhorting us to direct our attentions here, to focus on that over there, and to act in that or this way. People, in their bodies, experience objects and spaces not as point coordinates on a three-dimensional map, but dynamically and interactively.

The folk model of human cognition portrays the way people apprehend their surroundings as a sequential process, beginning with the "out there" of environmental perception, and then proceeding, when an environmental stimulus reaches the sensory systems, to the "in here" of cognition and interpretation, and finally resulting in an "out there" of action. It turns out, though, that the relationship of "out there" to "in here" is not a simple one of progression but a complex, intermeshed, and recursive one. Psychological research, neurophysiological evidence, and various kinds of brain imaging studies all show that the boundaries separating sensory perception from cognition and sensory cognition from motor action are at best indistinct and perhaps nonexistent. So when you are thinking something over, it may seem to you that all the action is inside your mind. But in reality your embodied mind already is engaging with the physical environment. It is quietly preparing an action plan for what it will or might do "out there."

People literally cannot think without a goal in mind. One prominent neuroscientist writes that we should think of the brain not as a machine for thinking but "essentially an organ for action." Sensory cognition, another contends, is "basically an implicit preparation to respond" to things in the world. This means that people, whether they recognize it or not, experience built environments by selectively focusing on the opportunities a given space or object or structure affords them. The founder of ecological psychology, J. J. Gibson, coined a useful concept in the 1970s to convey how people actually experience the environments they inhabit: he called our cognitive understanding of these opportunities *affordances*. Gibson's notion of affordance has to do with the properties of an object or the features of an environment that suggest to us how it is to be used. A doorway clearly asks us to walk through it. It's almost as if something about the space or the object or the structure speaks to us, signaling how we might engage with it. We experience most buildings, streetscapes, and landscapes in a way that is fundamentally embodied, determined less by its allocentric spatial or formal composition than by how we apprehend its affordances.

Architects often privilege the concept of "space" when they design, which they conceptualize as an abstraction, a geometrically homogeneous void in which objects are placed at identifiable coordinates—remember how Walter Gropius designed his model homes for the Weissenhofsiedlung in Stuttgart. But most of the time, when people encounter the voids and objects that constitute our built environments, they do not direct their attention on space per se. Instead, what people register nonconsciously and what they consciously choose to focus on are the experiential opportunities offered by a place's affordances. Take the example of a staircase. From the point of view of affordances, the critical aspects of a staircase are the ratio of my leg length to the height of the risers and the depth of its treads. I cannot see even the most elegantly composed staircase solely as an abstraction; inevitably and nonconsciously, I also consider the actions it enables me, actually or imaginatively, to take. Since any space contains multiple affordances, the environments we inhabit are anything but mute, homogeneous voids: they are vibrant, settings for life, filled with imagined and actual actors in motion—they are the *action settings* in which we live our lives.

Contrary to our internal sense of things, never do we experience the built environment either holistically or passively. Nor can we just stop to take it in. Instead, we assess the usefulness to us of its various components, all the while either acting or formulating goals in preparation to act. This is true even if the "action" we undertake is simply observation. What this means is that we, without being aware of it, experience a built environment *actively*, with our many senses and motor systems engaged (in other words, in ways that are both intersensory and cross-modal). To make sense of the spectacle *Cloud Gate* presents, for example, we actively engage it while contemplating what it has to teach us about our place in our body, the city, the world.

Not everything in our built environment primes us by activating our many senses along with our motor system. If it did, we would be in a

perpetual state of cognitive overload. Amid the constant bombardment of stimuli, the nonconscious brain chooses what to pay attention to. It does so by using selection principles that it has developed over its years of helping you navigate your world. One criterion determines an element's proximity to our body, which will help us decide how we interact with or engage it, imaginatively or actually. Another assesses its perceived usefulness to our goal. And the last, perhaps most important, appraises its attention-worthiness. A feature in the environment activates our systems of sensory and motor cognition *only* when and if you pay attention to it. Perceived proximity, perceived usefulness to goal execution, attentional focus: as we inhabit and navigate the built environment, we rely on these three factors to help us parse the things with which we wish to engage from the things we don't, won't, or can't.

Because of the embodied nature of cognition, and because humans are in a near-constant state of goal formation, the built environment is anything but static and inert. Perpetually, dynamically, actively, we are engaging with the places, spaces, and objects of our surroundings: buildings, parks, squares with or without sculptures, streets, all the time relating to them with our whole bodies, and with our egocentric and allocentric mind-sets, with all of our senses. And in that dance, we pay most attention to those elements that fall within our *periper- sonal* space, which differs from where elements fall relative to our body in metric space. We perceive objects in peripersonal space as, in one way or another, falling within the aegis of our actual or imagined reach. It is mostly in peripersonal—not metric—space that the events of our lives unfold; anything outside that area we conceptualize as *ex- trapersonal* space. Elements in the built environment that fall within our peripersonal space are judged relative to our understanding of our own body. So if the width of a window exceeds the reach of the span of

Stairs are made for walking, Itamaraty Palace (Oscar Niemeyer), Brasilia, Brazil our outstretched arms, its spread will feel expansive; if a door's height reaches far above our hand when we stretch it above our head, it will feel grand.

One of the aspects of *Cloud Gate* that makes it so enthralling is that it elides the boundary that divides the carnival of peripersonal space from the horizon of extrapersonal space. By ushering Chicago's skyline into our peripersonal space, Kapoor's sculpture—like magic!—effaces boundaries we weren't even aware we lived within. Normally as we walk around Millennium Park, we would devote a good deal of attention to affordances surrounding us: the bench where we intend to sit to eat lunch, the path on the other side of the park, which leads to the Art Institute of Chicago. In the ambit of *Cloud Gate*, our experience is different, because the sculpture simultaneously homogenizes and distorts the space of our body, the space around our body, and the space of the world. We in our body, along with the whole of Chicago's visible skyline, become peripersonal space: a three-dimensional world pulsating with potential actions, activating our imaginations, provoking an unfamiliar whorl of cognitions. Transforming our spatial relationship to our surroundings, *Cloud Gate* gathers skyline and strangers all into the peripersonal space of our own body, making of them all tools in a composition of our own making.

But objects needn't use mirrors and pedestals to enhance awareness of our embodiment through design. With the Glass Pavilion at the Toledo Museum of Art in Toledo, Ohio, Kazuyo Sejima, the lead architect in the Japan-based firm SANAA, wrapped curving floor-to-ceiling glass walls around spaces to variously articulate, obscure, and reflect the physical boundaries between corridors, exhibition and ancillary areas, and other walls, all the while maintaining the sweep of visual continuity. The Japanese-born artist Shushatku Arakawa, with his wife, Madeline Gins, similarly effaced the boundaries between peripersonal and extrapersonal space in their own Bioscleave House on Long Island,

Glass Pavilion, Toledo Museum of Art (Kazuyo Sejima/SANAA), Toledo, Ohio

New York. Sloping, playground-like textured concrete floors in brightly colored surfaces challenge inhabitants to constantly adjust their body position in order to regain or maintain their balance. The house heightens people's awareness of their bodies as objects on planes in space, activating our sensorimotor systems of proprioception and of kinesthesia, pressing into the peripersonal space surrounding our body.

Embodied Minds in Action: Texture, Surface, and Metaphors at the Museum at the Stream, Antwerp

Kapoor designed *Cloud Gate* as a public sculpture and a cityscape. Fine art, more often than not, is explicitly intended to create emotionally affecting, memorable experiences. Buildings do this too. Buildings never leave us cold: if they're not affecting us positively, there's a good chance that they're affecting us negatively. Willem Jan Neutelings and Michiel Riedijk's Museum at the Stream (aan de Stroom), or MAS, in Antwerp,

Belgium, and the Cathedral of Notre-Dame in Amiens, France, illustrate two ways that larger, more complex affairs such as public buildings can harness our human sensory faculties to sculpt a vibrant and enriching experience of place. In each instance, designers have used specific compositional elements and details to heighten a sensory-cognitive experience that advances the story that the building, by virtue of its functions, tells.

MAS is a nine-plus-story-high museum enveloped in red sandstone and glass and devoted to the distinguished maritime history of Antwerp: in the seventeenth century, the city's port became a fulcrum of the international economy. Often portrayed in dramatic pictures shot from its waterfront facade, MAS towers above the low-scale, mostly stucco-covered buildings populating this recently redeveloped section of Antwerp's downtown. A long, low, single-story bank of sandstone-covered spaces forms the museum's visual platform; housed inside are a café and gift shop, where huge skylights offer dynamically skewed views of the adjacent museum.

Approaching the museum proper, we see large passages of rippling glass, which simultaneously echo the stream's fluid movement while introducing the museum's maritime theme. These lilting passages also elicit thoughts of heavy curtains, the kind that drape across a stage. Water, theatricality, port: these are the chords in a developing theme, sparking our sense of curiosity. Whether we consciously think it or not, we are bearing witness to a performance under way. The three ranges of rippled glass are sandwiched between and alternate with four red-sandstone-faced boxes; opaque and fortress-like, these protect the museum's local treasures. The irregular shape of these stone boxes creates an impression of rotating, upward movement, which adds to the dynamism suggested by the facade's apparent rustling, owing to the reflections in its glass.

Proprioceptive challenges: Bioscleave House (Arakawa), Long Island, New York

Museum at the Stream (Neutelings Riedijk), Antwerp, Belgium

Dense, light. Solid, penetrable. Static, spiraling. MAS presents as a forcefully enigmatic object. What might be inside these glass-and-stone-covered boxes? As we parse this out, we draw on a database of previous experience. *Containers*, including buildings with their defined boundaries, typically demarcate *categories* of function—a glass is a con-

Window-curtains: Museum at the Stream, Antwerp

tainer for holding water; a cardboard box one that holds goods. This "categories are containers" schema we extend beyond such quotidian experiences into the built environment, and also beyond: just as we are *in* a building, when we are involved romantically with another person we are "*in* a relationship"—a state of contact with another person that has definable boundaries, as a glass that holds water or a museum that houses valuable objects. Further, because MAS presents two different kinds of containers, glass and stone ones, we infer that its glassed-in areas house one kind of function, and the sandstone-faced areas another. Since these containers are sandwiched together in a single building, we assume that their respective functions relate. Our initial prospect of the building has already suggested as much by introducing us to its various themes: water, theatricality, port.

Other embodied, intersensory metaphors already have come into play. Windows are transparent planes that you can look through, as through clear water. Yet windows usually don't ripple. In their rippling, these windows convey a sense of time passing, of *change* in blowing breezes and dancing light. In another transformation of the ordinary, draperies rustle in the wind, glass is inert; yet here, glass looks like draperies rustling in the wind. All these are allusions to embodied experience, with its tactile, auditory, and visual sensations of change over time. And because they activate our motor sense of physical engagement with the building, they are also cross-modal. Metaphors, we know, provoke us to mentally visualize, or sensorimotorically simulate the specific interactions with the world to which they refer. Neutelings Riedijk's design instigates thoughts of a number of embodied metaphors, all of which establish and reinforce our nonconscious, bodily engagement with MAS.

The overwhelming impression we get from MAS is that of a palpable tactility. Our human sense of touch includes a number of dimensions, such as texture, pliability, temperature, density, and vibration; MAS's design plays with most of them. The designers seem to have set out to materialize in three dimensions the interdependence of our visual system with our sense of touch, and the enmeshment of our various sensory faculties with our motor system. This is a building that insistently invites us to either touch it or imagine touching it. This it does in many ways: through visual cues, the choice of materials, the preparation of those materials for the process of building, the management of scale, and the deployment of ornament. Neutelings Riedijk selected a rich red-toned sandstone that has visual depth and a complexity that conveys its ancient qualities. The architects then amplified the tactile allure of the sandstone by specifying that the blocks be prepared or dressed by hand for installation. As a result, the stone facing retains an irregularity that bears the mark of human modeling, with passages so sensual that just a glance ignites our tactile imagination. And what these visual cues suggest, scale reinforces, since each stone block is measurable with the reach of an arm. Then Neutelings Riedijk transformed construction details into ornaments that further reinforce these tactile associations. Every third hand-dressed block is secured to the facade with metal bolts that are sculpted to resemble a single hand, which is also the city's historic symbol.

These hand-bolts unleash a cascade of tactile associations that reinforce the museum's major themes. Hands built this city. Hands cut that stone. Hands held these blocks in place, until the hand-tooled-bolts replaced them. Hands made this building, and hands made the things found inside of it. We cannot but think such thoughts in the presence of MAS: by design, it pulls us into a profoundly engaged experience with the building and the objects that it presents.

Standing in MAS's plaza, you could scarcely avoid apprehending its tactile properties. And though we think of touch as a single sense, our tactile system includes a number of different but related faculties that feed into a single cognition. When you run your hand along MAS's rough-cut stone surfaces or touch its smooth, silver "hands," your muscles and joints sense the stone's and the hand's density, weight, and vibration. Your skin registers the stone's rough surfaces and the silver sculpture's smoothness and warmth, activating memories of our past experiences with textured stone and metal. Just *seeing* MAS's rough-cut

Hand-dressed sandstone with hand-bolts on exterior, Museum at the Stream, Antwerp

stone sparks our memory of what such a surface feels like on our skin—
its texture, its likely temperature, its density or porosity.

Any number of common verbal locutions capture the especially palpable force of visual-tactile sensation in our everyday experience. If we
deliberately steer our unknowing colleagues toward a decision we wish

them to make, they are pliable, "putty in our hands"; if we believe ourselves approaching a cherished goal, we might think, "I can just *touch*" it; if a colleague brags about a love conquest, she might say that she "nabbed" her mate. Mental simulations of tactile experience pack a special kind of power. They feel immediate, because—whether we know it or not—they are prompting us to imagine, or simulate, in our own minds and bodies something of our actual responses to those surfaces. Of the many kinds of intersensory collaborations among our perceptual systems, that joining tactile and visual cognition is especially robust. We know from brain scans that tactile sensations stimulate areas of our visual and auditory cortices, and visual sensations stimulate areas of the auditory and somatosensory cortices.

This is but one of the many insights that laypeople as well as designers can learn from the cognitive revolution: surfaces, and therefore materials, profoundly affect our nonconscious and conscious cognitions about the built environment. One implication of this is that any surface that does not enhance our experience of it diminishes it. Because tactile impressions involve our own movement or imagined movement (running your hand across the facade, for example), they activate our sense of ourselves as wholly engaged in our environments—especially as we can control what we touch far better than we can control what we see.

Vision remains the undisputed king of the human system of sensory cognition. Our brains devote approximately as much processing power to what we see as to all the information coming from our other sensory faculties combined. Still, our experience of MAS illustrates that our eyes never take information in from the world without assistance from other sensory faculties as well as from associative schemas, metaphors, and other memories. Neutelings Riedijk's design springs from the premise that in architectural experience, what we see also activates the parts of our sensorimotor systems devoted to touch. The architects ensnare us into a full-body engagement with MAS by deploying design techniques that elicit our nonconscious responses to the building as a whole, and to the museum's exhibitions celebrating Antwerp's once-international status as an arbiter and clearinghouse for international trade.

Amiens Cathedral's Sensational Orchestra: "Feeling" the Presence of a Higher Being

In the eight-hundred-year-old Amiens Cathedral, the master masons exhibited far less concern with creating a palpable sense of embodied immediacy than with constructing a place that would spiritually transport anyone who entered it. Skillfully manipulating our wayfinding abilities through our senses of proprioception and hearing, the masons built one of the finest examples, and the largest intact one, of thirteenth-century French Gothic architecture, while exaggerating our impression of Amiens Cathedral as a splendid—indeed, extraordinary—house of God.

It takes fewer than ten minutes to traverse the urban plaza that introduces one of the greatest architectural experiences on earth. The western towers of Amiens's cool, blue-gray stone facade stretch 358 feet toward the sky, making you in your body feel small; instinctively, you take a few steps backward, hoping to benefit from the slight reduction in scale that distance provides to better apprehend this massive edifice whole. Light and shadows weave its sculpted facade into a tapestry animated by spires, griffins, trefoils, quatrefoils, applied colonnettes, and standing arrays of familiar saints, gazing beatifically upon tourist and supplicant alike. A tight tripartite composition, running both vertically and horizontally, organizes the monumental facade, and the theme of the Trinity introduced here is continued in the footprint, or plan, of the building. The three-part ground-level portal corresponds to

Cathedral of Notre-Dame, Amiens, France

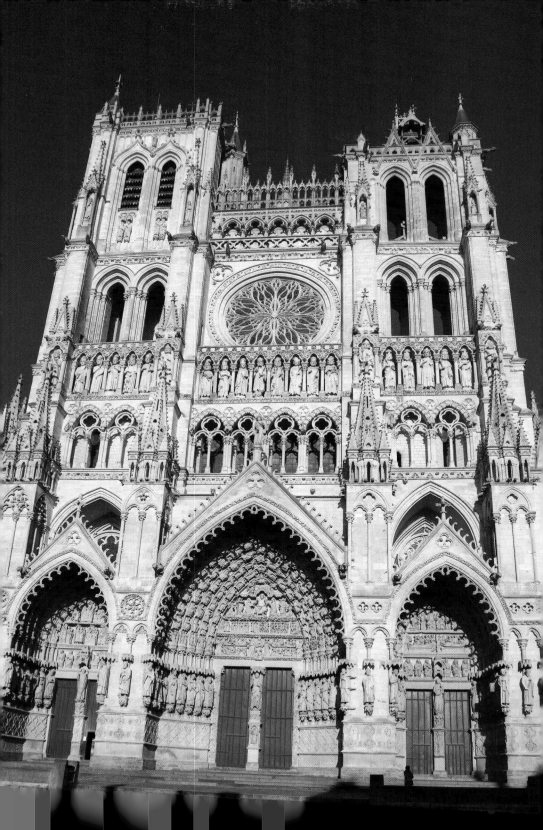

the organization of the interior spaces, a central nave framed by two interior side aisles. Sculpted figures with foliated capitals, saints huddled together, ease the transition in scale from the monumental facade to the entrance doors that beckon us to enter.

Once we are inside, an imposing universe spreads before us, enveloping us in a vast, shadowy nave of muted sounds, musty aromas of centuries-old lime and wooden pews, and light streaming from above. (Stained glass substantially darkened the light coming into the cathedral until World War II; Axis bombs shattered its medieval stained-glass windows, which were replaced with clear panes.) Seven massive paired piers lead our eye, columned pier by columned pier, into one small stretch of eternity, pacing our feet toward the soaring transept. Exaggerating the columns' breadth is the hand-sized measure of their slender, clustered colonnettes. Cold bites at your skin: too expensive to heat, the cathedral is cool year-round, and on subfreezing winter days you can see your breath. The nave's immense scale and the vertical line of the colonnettes also inexorably pull our eyes upward, toward the arched vaults and soaring ceiling, as we seek out the measure of the place. A straight axis directs our eyes and feet forward, toward the transept, but the nave's scale seems so immeasurable that we instead gravitate toward the church's lower-scaled side aisles, through which we slowly walk, musing occasionally on the statuary and illuminated candles on our way to the transept crossing and choir.

Rewind the tape. Now think of approaching, and then walking into and around Amiens Cathedral, but focus this time on sound. As you walk across the cathedral plaza, the sonic landscape, or soundscape, of Amiens, a midsize (~140,000) French city, rings and clatters with garbage collectors shouting and clanging metal-grated bins while their trucks idle on, auto engines droning, drivers honking, wary cyclists ringing their bells. Having reached the portal, you enter, and the moment the cathedral's portal door closes behind you, all these sounds are extinguished. You have arrived. A vast, airy expanse washes around you, where stone surfaces reverberate the hushed shuffling of wandering feet.

Constituting your experience of Amiens, and indeed of every place in the built environment, is not only what you see but what you hear: humans are exquisitely sensitive to sound. We differentiate the rustling of paper from the rustling of leaves. We distinguish rustling from rubbing, and scraping from scratching. The seemingly infinite world of sounds results from the combination of a constrained, definable set of sonic factors: intensity; frequency, or the rate of physical oscillation of the note sounded (high-frequency oscillations create "high" notes, and low-frequency oscillations create "low" ones); timbre, or the quality of its vibrations; duration, which is how long a sonic emission lasts over time before fading out. And since most sounds come not as a single emission but in multiples, released over time, rhythm and variation are often central features of sonic events.

Cathedrals create highly unusual sonic landscapes. In an interior as large as Amiens Cathedral, sounds linger; indeed, the length of time a sound takes to fade out approaches our experience of hearing a sound fade out when we stand in an open field. But in Amiens, unlike in an open field, its array of regular and irregular interior surfaces in the ambulatory, nave, side chapels, transept—absorb, reflect, and diffuse sonic emissions in a variety of ways. The building's materials, masonry and glass, reflect much more sound than they absorb. A large area contained by predominantly sound-reflecting materials creates Amiens Cathedral's distinctive soundscape, because in such spaces, sound waves reverberate for an unusually long time, from six to ten seconds. The result is what acousticians call *ubiquitous sound*, whereby the long reverberation of sound emissions makes it seem as though sounds come from everywhere and nowhere, offering us precious little information as to their sources.

The ubiquitous sound that we hear in these vast spaces differs from just about any other auditory experience in the built environment.

Above: Portal, Amiens Cathedral

Opposite: Interior, looking down nave,
Amiens Cathedral

Left: Interior, looking up at transept
crossing, Amiens Cathedral

Because our two ears are placed on opposite sides of our one head, sound waves hit them at slightly different times and with slightly different intensities. Like binocular vision, which facilitates depth perception, biacoustic hearing facilitates the ability to locate a sound in space in relation to our body; this is called *echolocation*. People combine echolocation with their sense of proprioception to mentally take stock of their environment and identify movements, which signify possible danger. Our experience inside Amiens Cathedral is not only distinctive but vaguely unsettling because there we stand, in a soaring nave that offers sight lines in every direction, but when we try to locate the sources of the sounds we hear—people's footsteps, the rustling pages of guidebooks, the dulcet tones of choral music—we cannot. Given that we are in a church, we know, logically, that we are safe. But the Amiens interior's ubiquitous soundscape forces us to relinquish some measure of control, surrendering it to something amorphous and ill-defined, as well as more powerful than we—something, in other words, akin to a higher being.

People also conceptualize time by relating it to their embodied spatial experience, speaking of life as a journey along a path and of unhappy episodes in their lives as dark moments or places. Smaller spaces seem to speed time up; larger spaces, by contrast, seem to slow time down. Because of its vastness, because of its distinctiveness and its powerful soundscape, to enter Amiens Cathedral is to give yourself over to some kind of liminal other time and space, which fills you with a sense of awe that transports you from your everyday thoughts and perhaps self-absorbed perseverations, concentrating your mind instead on your existential commonality with all of humanity, living and dead. That's why spending half an hour in a place like Amiens can seem like a full day well spent, and has the potential to become a day that transforms a life.

In human experience, design matters. We marvel at *Cloud Gate*'s transformation of our sense of ourselves in relation to Chicago's cityscape. At how MAS theatrically enmeshes us in the changing fortunes of human history; at how such an enormously light-filled, airy cathedral as Amiens could convincingly suffuse us with susurrant thoughts and emotions of transcendence. When we first encounter these magnificent places, we apprehend them mainly visually, as objects in space. But over time, even small passages of time, our experience of them is informed by all of our senses, and by our sensory and motor systems working in concert, and by the metaphors and memories we bring to and take from being there.

By design, *Cloud Gate*, MAS, and Amiens Cathedral all draw on and manipulate various aspects of embodied cognition to create distinctive experiences and places. They invigorate us, make us feel more alive, open up possibilities for us, and connect us with others. Good building and landscapes not only constitute but expand our horizons, challenging us to recognize and to contemplate the experiential possibilities of architectural and built environmental expression.

CHAPTER 4

Bodies Situated in Natural Worlds

Don't worry, the moon will rise
wherever you go. It does even here,
miles away from the land's monopoly
of red barns and bent grasses.

　　—MARY JO BANG, "How to Leave a Prairie"

The sky is a tarpaulin lashed: it leaves
no way for crawling out.
And even you are planted here,
who seemed to choose otherwise . . .

　　—ROSANNA WARREN, "A Cypress"

Many years ago, I spent three months traveling around the Indian subcontinent. From Dhaka, Bangladesh, I jostled on third-class trains and wedged my sweaty way into brightly painted buses, zigzagging through northern India, from Kolkata on the eastern coast to Mumbai on the west, detouring farther north for a brief sojourn in Nepal. Those months changed me and my life in countless explicable and inexplicable ways. One in particular I recall several times a month.

Early one morning a bus disgorged me in Delhi, where the old city's huddled, disintegrating streets and buildings became the datum for my days and memories—hot, dirty, colorful, and overrun with honking buses and cars, bicycle whistles, shouting people, crying and laughing children. On that first scorching day, spent navigating dust-covered streets and dirt alleys by rickshaw and on foot, I happened at sunset into the serene green expanse of a well-tended park, the Lodhi Gardens. My body flooded with relief: I'd escaped, escaped the hours upon assaultive hours of stress I hadn't even known I felt. Making my way into the gardens and up a small hillock, I was stopped, as if by the brute force of an angry man, by a small stone building, a monument.

MOHAMMED SHAH SAYYID'S eight-sided tomb, built in 1445, is like Kahn and Tyng's Trenton Bath House: that rare architectural composition that manages to be at once small and monumental. Something akin to an early Romanesque baptistery, its heavy, sloping piers buttress its corners to exaggerate the weight of its planar, masonry simplicity. Canted walls, compressed proportions, and opaque surfaces make Mohammed Shah's tomb look as though it erupted from the earth moments before. Steady, stolid, tense with energy.

Mohammed Shah's tomb, Lodhi Gardens, Delhi, India

It had become too dark to photograph, the sun almost wholly departed for the day. Searching for light, I glanced up to behold the stars in the handle of the Big Dipper glowing along with the moon in the sky. That same instant I recalled standing in a wide-open field half a world away, a child, exploring the fields on my family's farm in central Vermont. Watching the moon rise as the sun set; counting the one, two, three stars in the handle of the Big Dipper. And now here I was, all the way in India. Standing on an open, green plain, watching the moon rise as the sun set, counting one, two, three.

No matter where we live—no matter *when* we've lived, I thought in wonder—people have been searching for light, looking at this moon. Feeling relief as they step onto a soft green horizon, feeling the force that drives that green fuse. When confronted with the tomb's canted walls, bearing witness to this visible manifestation of the earth's elemental forces, to the gravity that pulls us down, down, always down, back onto the surface of the earth.

In the contemporary world, ever more people inhabit ever denser and primarily built environments that are, by definition, ever further removed from the natural world. Humanity's estrangement from nature is but the continuation of an age-old process of building, which originated less as a locus for economic self-improvement and more in the interest of survival. Safe, durable ready-made shelters in nature are scarce. So people build them. As larger aggregations of shelters became settlements, the more durable political, social, and cultural institutions of civilization took hold, requiring more construction. Eventually, settlements became large enough to require infrastructure, so humans constructed still more, buildings and aqueducts and bridges and sanitation systems, all to support their lives in built environments. Progressively, whenever and wherever financial means conjoined with new technological developments, highways, subways, parks, parking garages, supermarkets, and power plants followed.

The story of human history is a story of building, perpetually building to accommodate actual and imagined needs. Human-made environments fulfill functions and offer amenities that nature cannot, so we tend to conceptualize them as nature's antithesis—machines in gardens, standing tall against the ferocity and fickleness of the elements. But in truth, nature shapes how people experience the environments they build. Our human bodies—two eyes on our face, two arms at our sides, two feet on the ground, a head perched atop a gravity-defying skeleton—evolved over tens and hundreds of thousands of years by adapting to nature's forms and rhythms, by successfully confronting its challenges and seizing its opportunities, by discovering safety in its sheltered corners and possibility in its airy expanses. Our bodies flourish and perish on this and no other earth: we thinking, breathing, sentient creatures sleep when it's dark and amble in the light, drink and bathe in water and consume its animal and vegetable offerings until, when it's time, we dissolve to dust.

Our experience of ourselves is embodied—situated in our bodies—and these human bodies of ours are also situated in the foundational environment of the natural world. Nature's geography and physical elements radically shape human cognitive experience in myriad ways. Nature restores us. All we need to become aware of its salutary effects is to step outside and take a deep breath, or pass from a crowded sidewalk into a verdant park. Some of the profoundest ways that nature affects us take place elsewhere, outside our conscious awareness, as our bodies and brains respond biologically, neurochemically to nature's cornucopia of offerings. We may know that a dearth of fresh air or a paucity of greenery or a scarcity of natural light degrades the moods of people vulnerable to depression. But such factoids misrepresent how pervasively nature's presence or absence influences people's cognitions and emotions. For even if we do appreciate the importance of nature in the abstract, we tend not to apply that knowledge to our own experience, especially since a great deal of the time, most of us simultaneously pay little attention to our environments and enjoy an inflated sense of our own agency. In casting about for the reasons underlying a dark mood or an embarrassing senior moment, most people would be unlikely to consider the hours they'd just spent in a windowless room.

Human life in the natural world gave rise to and shaped the structures and capacities of our minds and bodies. Our long evolution in earth's varied habitats and ecosystems, each with its own climate, topography, and greenery, imbued us with sensitivities to and proclivities for certain environmental patterns and ways of being in the landscape. My sense of relief upon entering Delhi's Lodhi Gardens, which afforded me the opportunity to replenish my depleted cognitive resources, exemplifies one such legacy. Another is that people are drawn to enclosed areas (where we can take refuge) coupled with views of and access to open, expansive terrain where we can "prospect" for opportunities.

People Need Nature

Humans have inhabited the earth for between 200,000 and 450,000 years. Until the last 10,000 years or so, we dwelled in varied climates including, but not confined to, the temperate, grass-covered savannahs that covered swaths of sub-Saharan Africa. Nomadism slowly dissipated with the dawning of agriculture, which more or less coincided with the establishment of permanent settlements where people organized themselves into progressively larger and more complex social constellations. Many scholars date the earliest cities to between 4000 and 3000 BCE; ancient Uruk, one of humanity's earliest cities (in today's Iraq), contained between 50,000 and 80,000 inhabitants. Even with the birth of urban societies, though, for many thousands of years after, most people continued to inhabit overwhelmingly unbuilt, nonurban settings.

The large-scale urbanization that facilitates and accompanies modern economic development railroaded through local cultures, resources, and across Europe and eventually around the globe, in the last two hundred years. Today ever more of us dwell in metropolitan areas, yet for all but a tiny fraction of *Homo sapiens*'s time on earth, the environments we inhabited were dominated by nature's rhythms and patterns. During those previous hundreds of thousands of years, generation after generation of humans successfully negotiated nature's diversity, managed its challenges, and achieved sufficient enough mastery over its elements. From the standpoint of our evolution, modern cities, let alone megacities, are so recent that humans have not had time to biologically adapt.

Genetically we are predisposed to crave and take a singular kind of pleasure in environments where nature's presence is palpable. Even if systematic and individual variations exist in our proclivity for nature owing to personality, gender, age, and cultural upbringing—and they do—we embodied humans have evolved as a *biophilic* species; meaning that we are drawn to nature: we like to feel a connection to it in our homes, our offices, our communities. Our very genes are encoded to link our well-being—our *being* well and our *feeling* well—to sustaining an intimate connection with the natural world. This applies to urban dwellers and to country folk, to people in all kinds of environments, to people of every ethnicity.

Countless studies reveal our biologically grounded dependence on nature. Consider, for example, two adjacent, architecturally identical residential courtyards located in the same low-rise housing complex in Chicago. One, which we'll call the Green Courtyard, contained plantings, grass, and trees. The other, the Gray Courtyard, was paved with concrete. Same city. Same neighborhood. Same building design. Residents of similar socioeconomic status and background. Yet the lives of the two sets of residents—especially the children—differed depending upon which one they called home, with residents of the Green Courtyard buildings markedly healthier, physically and psychologically.

Green Courtyarders coped better with stress. They better managed interpersonal conflict. Most astonishing, the children exhibited superior overall cognitive functioning. Dozens of subsequent studies confirm these findings, including work in recent years on communities in Baltimore, Chicago, Philadelphia, and Youngstown, Ohio, correlating significantly reduced incidences of crime (both property and violent) to increased greenery in public places.

At least one reason why regular access to nature reduces crime rates and stress is that it improves people's cognitive faculties. We know that people's ability to concentrate with focused effort when they need to is a critical faculty that facilitates so much of our ability to think clearly and effectively. We also know that this ability is easily depleted. According to environmental psychologists Rachel and Stephen Kaplan, enjoying a natural landscape replenishes our attentional resources effectively by promoting what they call effortless focus. Natural environments engage

Residents of housing projects thrive when nature is visible and accessible, and don't when it isn't: Ida B. Wells Housing, Chicago, Illinois [demolished]

our curiosity and attention without willful determination on our part.

The greater a city dweller's access to greenery, light, and open spaces, the better she will solve problems and understand and take in new information; the more mastery she has over her limited attentional resources, directing and sustaining them where and when she wants; and the more effective she will be in regulating her emotions. All this translates into improved psychological well-being and better interpersonal relations. And there's more. Residents lucky enough to live in a housing development surrounded by vegetation—trees, grass, flowers—enjoy and maintain stronger social ties with their neighbors; they enjoy a sense of community more robust than residents in similar buildings lacking in these natural features. Green Courtyard residents also believed that their environs were safer than Gray Courtyard residents did, and the crime statistics data have consistently supported their perceptions.

Building cities that neglect our human need for nature strains public resources and exacts a high cost on everyone. Yet legions of affordable housing developments—ones already built and ones still on the drawing boards—ignore or pay lip service, at best, to these fundamental human needs. For this reason and for many other reasons, all but a few of this country's affordable housing developments disadvantage the very people they purport to help. Exceptions like the green-roofed, multi-terraced Via Verde in the Bronx, developed by Jonathan Rose Companies, and designed by Grimshaw Architects in collaboration with Dattner Architects, only seem to prove the rule.

Human *biophilia,* or love of nature, influences not only our built environmental experience in the immediate moment but also in our memories. Nature's presence or absence affects how we remember where we've been and therefore who we are. Recalling that people's autobiographical memories are processed in the same part of the brain involved in cognitive mapping—autobiographical memories come

Affordable, sustainable, green, humane: Via Verde Housing (Grimshaw and Dattner), Bronx, New York

packaged by place—means that our experiences with nature as children play a significant role in our sense of self and identity. Think about this in concrete terms. People remember the neighborhood in which they grew up more fondly if it offered meaningful access to nature. Part of what made your friend's childhood a happy one was the easy access he had to a nearby park or the foliated view from his bedroom.

Predictably, people's biophilic preferences extend beyond the home.

Whether we are toiling away at the office or working out at the health club, more access to nature better nurtures our well-being. Herman Miller, an office furniture company with a base in Zeeland, Michigan, confirmed this after it moved its employees from an outdated manufacturing facility into a new building designed by William McDonough + Partners which they called the "GreenHouse," outfitted with courtyards, internal gardens, and skylights. Daylight and greenery graced internal spaces, including corridors. Within six months, employees' workplace satisfaction and performance measurably improved. After nine months, their productivity had increased by an astonishing 20 percent. Employees found themselves healthier, less distractible, more relaxed, and more highly motivated to work. In another study, merely substituting the standard levels of artificial ventilation used in most office environments with a level approximating a naturally ventilated space dramatically improved workers' overall cognitive performance. Making workplace environments that simulate or evoke natural conditions, including natural light, positively impacts a company's vitality; conversely, failing to do so exacts a cost, not only on employees' sense of well-being but also on their health and productivity—the economic bottom line.

We have seen that placing a postsurgical patient in a hospital room with a pastoral view, rather than in one facing a brick wall, will result in a speedier and less painful recovery. When inpatients spend time in what hospital administrators call "healing gardens," their heart rates slow and their cortisol and stress levels fall. Such effects transpire with astonishing rapidity: patients—indeed, even people unafflicted by medical maladies—become aware of them after three to five minutes. And these salutary physiological effects are measurable within just *twenty*

In healing gardens, heart rates settle and cortisol levels fall within minutes: Crown Sky Garden, Lurie Children's Hospital (Mikyoung Kim Design), Chicago, Illinois

Workplaces with ample natural light boost job satisfaction: conference room, Scottish Parliament (Enric Miralles/EMBT Architects), Edinburgh, Scotland

seconds of such exposure. What makes for good healing also makes for good workplaces, good schools, and good homes. Designs that offer access to nature or simulate its greenery, climate, and topography affect us beneficially for the simple reason that people thrive in environments where nature continues to nourish our well-being.

NATURE BEGINS WITH LIGHT

Among nature's elemental features, sunlight is its most celebrated. Light. Daylight. Let there be light. Flooded with light, washing light, *isn't the light beautiful* light. *God saw the light, that it was good.* Humans revere, indeed worship the light of the sun, and probably have since the beginning of our time on earth. And in built environments, natural light confers upon humans a plethora of salutary effects. Natural light remains qualitatively and quantitatively different from artificial light; it is hundreds of times brighter and more complex in spectral hue. Humans consistently prefer natural light, from which our bodies derive physiological and psychological benefits. Daylight delivers the sensible experience of warmth (or can); suppresses the body's release of melatonin, the hormone that coddles us to sleep; nourishes our body with the vitamin D that buttresses our immune system and stimulates bone growth and strength. No one really needs evidence to establish people's overwhelming reliance on and preference for natural light, but here it is: given a choice, people will consistently choose to spend their time in rooms where lumen levels approximate those that prompt the body to suppress the release of melatonin.

Even workplaces that offer more indirect than direct access to nature, but admit ample natural light, as through views and skylights in a conference room, report of higher levels of job satisfaction. Retail environments that cater to people's innate preferences for daylight attract more customers and retain them longer. When one supermarket, which had been housed in the standard-issue windowless box, moved to a building outfitted with multiple skylights casting abundant natural light indoors, its sales increased by nearly 40 percent. Natural light, like pastoral views, heals the sick and improves the well-being even of the well by affecting our cognitive processes in profound, though sometimes subtle ways. That hospital inpatients placed in naturally illuminated rooms (even absent a view) sleep better and enjoy more regular circadian rhythms than their counterparts in artificially illuminated rooms is perhaps predictable. More surprising are the additional healthy effects of daylit rooms. Patients feel less stressed. They report feeling less pain and heal more quickly. They suffer lower rates of mortality. Subjective impressions are substantiated by measurable data.

Children in properly daylit classrooms focus better, retain information better, behave better, and score better on tests. Daylight measurably affects people's moods (provided that glare and temperature are properly controlled) and is a proven palliative that alleviates the symptoms of mental illness, especially in people suffering from bipolar disorder and seasonal affective disorder. If daylight helps heal the sick, will it not also improve a sense of emotional balance in anyone?

Views and paths invite us to explore unknown places: Connecticut Water Treatment Facility (Michael Van Valkenburgh Associates, building by Steven Holl)

Natural light, a boon to human physical and mental health, also smooths the way to easier social interactions. People don't tend to notice their physical environments much anyway, but they do so even less when interacting with other people. Still, our nonconscious receptivity to natural light and its soothing properties persists. One study divided subjects into two groups and placed them in an identical social situation, with one group in a brighter room (1,000 lux of white light), the other in rooms with varying degrees of dimmer light. The subjects in the more brightly lit room quarreled less than their counterparts in the darker room.

We are so biologically wired to embrace the natural world that, in addition to greenery and light, we respond strongly to natural materials, biomorphic forms, and specific topographical features. These include ones that were critical to the African savannahs, where our ancestors thrived for tens of thousands of years: gently rolling hills, even ground cover, meandering pathways, and copses of trees and shrubbery that screen illuminated clearings. Think of New York's City's Central Park, or indeed any park, including this one surrounding a water treatment plant in Connecticut. We comprehend the overall structure of its "prospect and refuge" landscapes in one sweeping view, surveying the enticing places they offer for our exploration. Without even thinking about it, we can identify the areas where we can take refuge to hide and to rest, as well as to prospect so that we can find water and food. Whether for reasons of evolutionary adaptation or pragmatism, people gravitate toward "prospect and refuge" landscapes. They seek them out in real life, and they like them even once removed, in landscape paintings by J. M. W. Turner, Albert Bierstadt, and Sunday painters; in photographs by Ansel Adams and a thousand commercial studios. Exposing people even to *representations* of natural landscapes improves people's physical and mental health. So if your doctor or your boss can't cut a window through the office wall, then hanging a picture of nature or getting furniture with biomorphic forms and natural materials helps more than doing nothing at all.

Managing Human Experience through Form-Making: Louis Kahn's "Deep Reverence for the Nature of Nature" at the Salk Institute

Of course embracing a site's greenery, topography, and light, and eliciting "prospect and refuge" behavior entails more than cutting skylights in ceilings and plunking down potted plants in corridors. To explore the range of ways that the natural world can inform a project's design, we can visit one of modern architecture's greatest and most beloved icons, the Salk Institute for Biological Studies in La Jolla, California, by Louis Kahn. Jonas Salk, the client for the eponymously named institute and the developer of the polio vaccine, believed that major breakthroughs in scientific research necessitated both the rigor of scientific method and the freedom of creativity, and he worked closely with Kahn to bring to fruition a complex of research laboratories and private offices sited on the crest of a sandy cliff overlooking the Pacific Ocean. The Salk Institute (altered in 1996 by a much-needed, though grievously banal, addition) deliberately appeals to people's inherent biophilia in obvious and less than obvious ways. Kahn gracefully integrated the complex into the existing site and invoked schemas of "prospect and refuge," introducing different aspects of our human connection to nature—our human nature—in carefully sequenced stages. The result is an enthralling architectural experience which synthesizes moment-by-moment experience with evocations of nature's enduring infinitude.

We come upon the Salk Institute in one of two possible ways: from the south (an approach mirrored on the north, but rarely used) and from the east. From the south, our first glimpse of the building, across a grassy knoll, presents a blank concrete monolith of a wall, punctuated by four projecting concrete prisms, each housing the deep shadow of a small entrance. It's a bit like stumbling upon the walls of a medieval

What's behind that concrete monolith of a wall? South facade, Salk Institute for Biological Studies (Louis Kahn), La Jolla, California

fort, simultaneously forbidding and intriguing. We cannot but wonder what's *behind* that wall. Can I go there? What would I see if I did?

If we had approached the original, pre-addition Salk Institute from the east instead, we'd start in the parking lot located off the busy Torrey Pines Road. Our first glimpse of the complex was through the foliated screen of a grove of eucalyptus trees (since cut down to make way for the addition), a buffer zone between the multilane road we've just left and the institute proper. Already, as we emerged from our car, we had entered nature's world, with the building discernible only by the edges of its symmetrically receding blocks. As in the south approach, from this first glimpse we deduce that this is no ordinary building. Instead of the conventional building-site formula—large object plunked on ground—the architecture here is merely a three-dimensional frame, its seeming function to introduce the stunning infinitude of the Pacific Ocean and its sky. Nature's awe-inspiring presence commands

Kahn designed the Salk Institute with "a deep reverence for the nature of nature": view from Pacific Ocean, Salk Institute

Original east entrance through a screen of eucalyptus trees, Salk Institute

our attention. The symmetry of the two laboratory blocks serves the same function as the A:B:A:B pattern on the south facade: architecture quietly reassures us of human presence. The Salk's restrained, symmetrically placed laboratories are slung low, hugging the cliff, framing the horizon, so that wherever we start from, nature dominates. These buildings do not stamp their feet and wave their arms; no "look at me!" moments here. Instead, the relationship of the site to the size, simple volumes, and restrained materials of the buildings beckons us to prospect, to explore.

These initial views of the original Salk Institute offer up easily comprehensible images and patterns because they are attuned to the mechanisms and—especially—the limitations of human visual cognition. The visual field in which human eyes perceive things as sharply etched, called the fovea, constitutes a mere two degrees in extent, a region so tiny that it can be completely obscured by your thumbnail if you hold it approximately fourteen inches from your eye. Because our face and feet are oriented in a direction we call "forward," to see what's behind or even at a sixty-degree angle from that focal point in front of us, we must turn our heads, our bodies, or both. Outside that two-degree cone, the resolution of our vision becomes astonishingly poor, though you may not be aware of this because your brain, using details gleaned from rapid scanning and based on memories of past scenes, supplies information to fill in the details your eye fails to capture. At any moment, much of what a person thinks he sees of the world in his peripheral vision—patterns, rhythms, and general compositional elements—is little more than an imaginative filling in of the blanks with the wonders of meaning derived from fleeting glances. As a result, human sight excels at *rapid gist extraction,* our efficient ability to extract essential visual information from our environments so quickly (twenty milliseconds) that the speed is literally within the blink of an eye, and when we do extract a scene's gist, we are rewarded with a little jolt of neurotransmitters that give us a sense of pleasure. The Salk complex defers to its duneside site while conveying a sense of order that is obviously human-made through the repetition of its simple rectangular volumes on the south and north facades, and through the symmetry of the laboratory blocks on the east facade. By designing the complex to both meld into the site and create a sense of mystery as to its identity, Kahn skillfully manages our initial emotional response to this place. It is as though he is saying directly to us, Forget the road. Here, you enter an oceanside refuge, what he called "a world within a world."

Even as we catch sight of the Salk Institute's centrally placed plaza,

Directing our eyes to the horizon and minimizing distractions: Central plaza with fountain, Salk Institute

it's not the forms of the architecture that command our attention. In the symmetrically arranged pair of light gray concrete volumes, Kahn minimizes the distractions that interrupt the sweep of our eyes toward the visual center of the composition, the Pacific's glistening horizon. Distracting views of staircases, windows, corridors, doors—ordinary architectural indicators of human presence and movement—are mostly suppressed. These approaches insist upon the architecture's *unobtrusiveness*, as Kahn doggedly designs our attention *away* from the buildings, redirecting our gaze on to the light-drenched, wind-swept site, to the dark horizontal line of the Pacific, to the clean grandeur of La Jolla's blue skies.

Humans are wired to scan for changes and anomalies in their immediate environment. Kahn breaks the quietude of our initial ingress, at first, with the sound: water gurgles into the channel embedded into the plaza's travertine pavement, with the fountain that feeds it emitting far more noise than its small size might suggest. Time seems arrested in everything except the splashing of water from a spigot that we cannot see. The channel fountain, which recalls those at the Alhambra in Granada, Spain, and in Indian Mughal mausoleum-gardens, measures only slightly wider than a human foot. Our auditory and proprioceptive faculties are put on alert: hearing the water's gushing, intrigued by this watery "line of light," we are determined to enter the plaza immediately. Is that water in the channel moving? Is it cold? Will my foot fit in? Kahn designed these approach sequences to slow us down and help train our focus on the substantive essence of the Salk Institute's mission: biological research, which is nothing less than an inquiry into the profound mysteries of nature. This he does by skillfully managing our visual, auditory, and proprioceptive cognitions to repeatedly redirect our attention to the site's natural surroundings—the eucalyptus trees, the crest side view of the ocean, the warm sunlight reflecting off the white, empty travertine. That's why Kahn, when speaking to an audience, told them that he had designed the Salk Institute "out of a deep reverence for the nature of nature. Built into us," he continued, "is a reverence for the elements, for water, for light, for air—a deep reverence for the animal world and the natural world."

Human bodies and minds evolved also with some elemental forms drawn from nature that are not embedded in specific geographies of place. This includes certain essential shapes and compositional patterns, their interplay with earthbound materials, and the effects of gravity on them. When people encounter a building, cityscape, or landscape, rapid gist extraction helps us to create a mental image of a scene's rough organization—"rough" because our perception cannot offer a *precise* account of what's on the ground. Our mental representations of our environs are only as accurate as we need them to be. Pattern, coherence, regularity, and contrast prevail over accuracy and precision; schemas and mental representations fill in the rest. Because of the human eye's limitations (especially in comparison with other animals), the way people see might better be described, writes one psychologist, as "solving ill-posed problems by adding assumptions about the world."

Scanning the Salk Institute, we rely on form-based cues such as edges, angles, corners, contours, and curves, testing them against an internalized storehouse of basic compositional patterns and volumes. (One of these patterns is the reassuringly predictable bilateral symmetry, which guides the overall arrangement of the Salk's structures when approached from the east, and which we discuss further in chapter 6.) The largest and simplest set of compositional schemas is grounded in a storehouse of basic figures known as geons. Geons, in the words of the vision scientist who discovered them, Irving Biederman, are "viewpoint-invariant," meaning that we discern their forms, individually and in combination, regardless of where we physically position ourselves in relation to them. That's why we don't need to change our position relative to one of the Salk's exterior laboratory staircases to know that it, and all the others, are rectangular prisms with parallel planes and edges.

We identify geons using their edge configurations—straight or curved, parallel or intersecting. Fewer than forty figures compose the full set of our internal storehouse of geonic figures, the irreducible elements of what

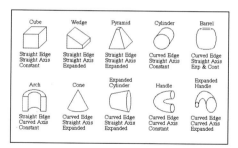

Some geons and their variants

Biederman calls our "recognition by component" way of seeing. Given the complexity of our visual world, forty may not sound like a lot. But because each geon has several variants and can be joined to any other, at any scale and in any combination, forty or so geons suffice to help us make sense of nearly everything in our visual world: any pair of geons can be combined to create over 10 million possible shapes; any three geon variants make over 300 billion possible shapes, and so on. Geons facilitate our rapid comprehension of the myriad form-based cues that the world throws our way—even if, in actual experience, some creations make geons easier to identify than others.

The universality of this shared mental storehouse rests in the regularity of geonic forms owing to their configuration by atomic forces and gravity. Geonic shapes abide by the principles of the physics of matter. Rectangular prisms, for example, pervade the material world because to make a volume stable, we must do nothing more than place planes and edges parallel or perpendicular to one another. Even not formally educated, indigenous children living in the Amazon, isolated from other communities, understand these basic geometric principles—indeed, they understand them just as well as most middle schoolers in the United States. Just through our interactions with the objects of the world, our internalized storehouse of geons gets reinforced and confirmed. Touching, even just seeing an object—a ball, a book, a teakettle—is enough for us to infer its overall shape, and then apply that

Identifying geons: triangular and rectangular prisms (compare with Heydar Aliyev): Ypenburg Housing (MVRDV), the Hague, Netherlands

Searching for geons (compare with Ypenburg Housing): Heydar Aliyev (Zaha Hadid), Baku, Azerbaijan

inference to the visual stimulus that is actually before us. In short, our cognitive reliance on geons suggests that the Platonic solids and principles of Euclidean geometry in the built world resonate with the frameworks our visual systems employ to help us see.

Managing Human Emotions through Materials, Textures, and Details

In the next part of our sequence into the Salk Institute, namely our longer exploration of its central plaza, Kahn lets the spectacle of nature's monumentality fade away. He mainly leaves behind the obvious use of reassuringly simple geonic forms and the geographically contingent elements of nature—greenery, topography, light—to orchestrate a more conventionally architectural experience. For the buildings of the central plaza, laboratories introduced by the staircase-office blocks, Kahn focuses the design around surface materials and their interaction with gravity. He deploys these elemental features of the natural world to ensnare us into an engaged, indeed interactive physical relationship with these buildings and—by extension—with the institution they embody and house.

To return to our path into the Salk's central plaza: having inferred enough from what we've seen to identify where we should go next, we make our way into the plaza's center. As we do so, the importance of the site's topography and the overall composition of the buildings diminishes. Increasingly, its materials and surface details command our attention. We've seen that geons are one of the principal means by which we comprehend forms. When our brain identifies a scene and discerns its form-based cues, such as shapes and their orientation, the sizes and combinations of its geons, that analysis runs through our parietal lobe, which has the primary responsibility for integrating sensory information from various parts of our body and is where the homunculus is located. This pathway suggests that to understand form-based cues we need not refer to our memories of past experiences with similar forms.

How the brain analyzes the cues it gleans from surfaces is different. In order for us to make sense of surface-based cues such as texture, density, color, pattern, and so on, our visual impressions are primarily processed through a pathway that involves the medial temporal lobe and the hippocampus, necessitating that—in contrast to form-perception—we call up our memories of prior experiences with similar surfaces. Such memories will draw up a lot of other varied information, not only from vision but also from our emotions and from other sensory faculties—tactile sensations, smells, sounds, and more. Our responses to surfaces, consequently, are more likely to powerfully contribute to our holistic experience of place than our responses to forms. In short: form has wrongly been crowned king, because form-based cues elicit less of a whole-body, intersensory, and emotional response than surface-based cues do. Surfaces we experience emotionally and *palpably*. And in our societal discussions of the built environment, surfaces and materials are often not a major theme; in its construction, rich and enlivening materials are deemed a luxury, often "value-engineered" out of existence.

Materials, patterns, textures, and colors shape our lasting impressions of a place as profoundly as does its overall formal organization and composition. Listen, now, to Richard Neutra, who designed some of the twentieth century's most richly elegant steel-and-glass residences. Describing the surfaces he recalled from his childhood, Neutra confessed, "Strange as it may seem, my first impressions of architecture were largely gustatory. I licked the blotter-like wallpaper adjoining my bed pillow, and the polished brass hardware of my toy cupboard. It must have been there and then that I developed an unconscious preference for flawlessly smooth surfaces that would stand the tongue test, the most exacting of tactile investigations."

Neutra's early mentor and employer, Frank Lloyd Wright, also concentrated much of his design energies on surface-based cues. Often he coupled highly textured surfaces with smooth ones—most famously, in the living room at Fallingwater in Bear Run, Pennsylvania. Wright sometimes used hand-hewn and natural materials and textures to stimulate not only our visual and tactile systems but also our proprioceptive system. In his design for the majestic Imperial Hotel in Tokyo, he paved the pathways outside the entrance with unpolished lava stones that were native to the region so that guests approaching the hotel needed to pay attention to the walkway's uneven surfaces. As Wright ushered guests toward and eventually into the hotel lobby, he made sure that the lava stones were polished to progressively higher states of finish, so that by the time a guest reached the front desk, she tread on smooth, well-polished surfaces. The Imperial Hotel's guests may or may not have consciously registered this subtle variation in the surface paving, but many probably reached the registration desk with a palpable sense of relief. Now I can relax.

Richly textured materials and surfaces—like the lava stones at Wright's Imperial Hotel, or the Salk Institute's travertine, concrete, and teak—elbow their way into our peripersonal universe by eliciting multisensory, emotionally rich nonconscious and conscious cognitions. Take the teak panels in the Salk's staircase-office blocks. People like wood. They are drawn to it for countless reasons. In comparison with metal, wood maintains a fairly constant temperature. Coloristically, wood skews warm, in hues of reddish-orange browns, a palette people tend to find appealing and subtly stimulating. Wood's grain exhibits an appealing tension of pattern and irregularity. Because wood commonly appears in residential architecture, it simultaneously elicits associations of nature on the one hand, and domesticity on the other. Travertine also elicits a rich associative trove, echoing some of what we glean from wood (nature, incident, texture) while also evoking an almost pathos-filled coupling of hard permanence with porous fragility and the creamy, rich, pockmarked stone of ancient Rome.

When a building's surfaces advertise the traces of their construction, they elicit our palpable, intersensory engagement in another way as well: by offering us opportunities to mentally simulate the process of their making. We saw this dynamic in the hand-dressed stone surfaces of Neutelings Riedijk's Museum at the Stream in Antwerp. Neutra, pondering his own reactions to the surfaces of handmade objects, explained its mechanism. He wrote: "Viewing hand-formed pottery, or the lines of a draftsman, or the lettering of a calligraphist, we unconsciously identify ourselves with their makers: We seem to follow vicariously the imagined muscular exertion in the nervous experience of the craftsman, as if experiencing it ourselves." Neutra's hypothesis, developed decades before the research tools were available to substantiate it, has been all but confirmed by the discovery of the brain's system of *canonical neurons* and *mirror neurons*.

Canonical neurons control motor actions; located in the brain's frontal and parietal lobes, they fire when we are doing something such as hand-building or throwing a clay pot, and they also fire when we do nothing more than look at an inanimate object, like a lump of clay, that we imagine ourselves manipulating with a goal in mind. Mirror neurons (also located in the frontal and parietal lobes) also fire when we execute a given action such as sculpting clay and when we mentally simulate that action; they also fire when we observe *someone else* executing that action. The brain's canonical and mirror neuron mechanisms indicate that in our experience of built environments, obviously human-made surfaces as well as manipulable objects really do prompt us to simulate the process by which they were crafted.

The operations of canonical and mirror neurons help to explain the visceral power of our responses to both form and surface-based cues. When we look at an object with which we might potentially engage or mentally prepare to undertake a given action, such as opening a window or ascending a flight of stairs, canonical neurons fire. Mirror neurons fire not only when we prepare to open a window or walk upstairs, but—amazingly—when we merely watch another person performing such an action, as though, in order to understand what that person intends to do, we imagine ourselves performing the same act. So these neurons "mirror" the actions of the person we observe. The discovery of the canonical and mirror neuron mechanisms supports the emerging cognitive neuroscientific view that the human motor system may not be distinct from our sensory faculties, and that they may be two components in a single, unified system. Perception is never passive. Perception is perception for action.

If we pay attention to any object or element—a staircase, a ramp—which we associate with a given action—ascending a staircase step by step or pacing up a ramp—our mirror and canonical neurons can fire. This, in addition to the visual dynamism of combining diagonals with spiraling lines, explains why we see Le Corbusier's famous coupling of a staircase and a ramp in the Villa Savoye as so dynamic. Looking at these two means of ascent or descent, we might nonconsciously feel a faint sense of activation in our legs and torso.

Scientists also continue to unearth our complex psychological and neurological responses to surface-based cues such as materials, textures, colors, pliability, and density. Exploring the canonical and mirror neuron responses to textures and materials is a rich area for future research. In the meantime, we already have a trove of studies illustrating the psychological power of our responses to surface-based cues. One we discussed earlier: students participate more in group discussions if cushioned furniture, throw pillows, and rugs bedeck their classroom. Studies in social cognition, which probe the judgments that people make about others and the decisions that they make

Perception is perception for action: two means of ascent at the Villa Savoye (Le Corbusier), Poissy, France

about their own conduct, yield surprising results. For example, when a person holds a hot cup of coffee, he will be more likely to assess a stranger as generous and friendly than if he holds a glass of iced coffee. A student will be a tougher negotiator seated in a hard-surfaced chair than in a cushioned chair. Meet a new person while touching a rough texture and you're more likely to recall the exchange as "rough"; meet someone while holding a hard object and you'd be likelier to perceive her as "rigid." Such data suggest that people metaphorically extend the schemas they construct of experiences with surfaces—we *see* a rough texture and know that it will *feel* rough on our fingertips—into arenas of our lives that fall far from the embodied experiences in which they originate.

Surfaces and forms come together, of course, and one aspect of a surface that integrally relates to the form it envelops is its shaping by the forces of gravity and the physics of matter. Mostly by living on the earth we understand gravity's basic principles. We can infer the approximate trajectory of a baseball once the batter hits it in the air. We know that standing on a vertical axis perpendicular to a flat plane will hold us steady, and that if we compare our relatively unencumbered heads "up there" to the load-bearing, spreading feet "down there," we will feel in our bodies the compressive forces of gravity moving downward. Similarly, we understand and even *feel what we see* when a designer articulates the force of gravity on his building, as did the designer of Mohammed Shah's proud tomb, and as Daniel Burnham did in the flaring base, with six-foot-thick walls, of his seventeen-story Monadnock Building in Chicago. A certain unease—or to put positively, dynamism—results too when the forces of gravity are contravened: Niemeyer deliberately challenges us in his unbuilt Museum of Modern Art in Caracas by inverting a pyramid to perch it atop a rocky cliff.

As humans grow from infancy to adulthood, we acquire and internalize a basic knowledge of the essential principles of gravity and the physics of matter. Such understanding helps us to know—sometimes without even knowing that we know—that two like-sized objects can differ in weight and would require different levels of physical exertion to move them. Embodied knowledge helps us intuit that a heavy object suspended above us, such as part of a sculpture or a building, might fall, potentially on us. People acquire a vast body of knowledge simply by living embodied in the world, as an object among objects, and as matter in space.

Teasing gravity: unbuilt project for the Museum of Modern Art (Oscar Niemeyer), Caracas, Venezuela

Taunting gravity: cantilever at CCTV Headquarters (Rem Koolhaas/Office of Metropolitan Architecture), Beijing, China

Perception for Action: Surfaces Activate Imagination

When we left off our approach to the Salk Institute, our ears were attuned to the gurgling water feeding its fountain. Ever attuned to changes in the surroundings, we instantly turn our attentions to the movement and sound of the channel fountain's water. As we make our way into the plaza proper, the water piques our curiosity; from there, Kahn orchestrates a composition that draws us in and encourages us to explore it. The plaza's linear channel of water that stretches toward the horizon reads not as a line on a plane, some abstract geometric composition, but as a vehicle bearing embodied meaning. In our human experience of space, lines mark paths, define boundaries, and articulate the edges within and between objects, materials, and spaces. Our eyes follow the path of this watery trough—which the scientists who work at the Salk call the "line of light," aptly capturing the channel fountain's dynamism, its glistening flow—to the vanishing point in the Pacific's horizon; immediately thereafter, we imagine our feet pacing out that path, walking alongside its line, heeding its directional call. Finally, our feet follow the path that our eyes staked out. So just the sight of this channel fountain initiates a succession of responses that encapsulate how we come to "feel" experiences in our physical surroundings. Our nonconscious perceptions and sensory faculties work together, intersensorily, and these all collaborate with our imagined motor responses. Little wonder that the artist Paul Klee described the act of making a drawing as "taking a line for a walk."

To reach the center of the plaza, we must descend several steps so shallow that we might barely register their existence, especially as we are focusing instead on other visible and audible enticements. At the steps' terminus, a bench running nearly the entire width of the plaza impedes our forward progress; to circumambulate it we must deflect our axis, which in turn shifts our perspective of the adjacent office-tower blocks.

Now we see them at an oblique angle. Quite suddenly, our mental image of this entire complex—a static, symmetrical arrangement of geonic prisms framing the horizon—disassembles before our scanning eyes. Those initially blank, monolithic concrete prisms pull apart, open up. Now the facades break into a porous rhythm of shadowed apertures and lightly incised planes. More: rather than sitting heavily on the ground, the staircase-office blocks seem perched lightly on top of it.

The walls of these staircase-office blocks resemble post-and-lintel structures that contain small private offices arrayed around an open-air dogleg staircase. Initially the Salk complex laboratories seemed like heavy, load-bearing monoliths slung low along the coastline; now those same buildings present as vertically oriented blocks, rising tall. In contrast to the complex's initially symmetrical, easily apprehended arrangement, now the floors of these towers stack in irregular dimensions. A tall bottom floor supports two shorter intermediate floors. An especially stretched top story caps the composition, its extra heft and height seeming to press these staircase-office blocks down, anchoring them into the ground. The stretched top story serves a function similar to the capital on a column, or a pediment atop a frieze.

In the staircase-office blocks, an intricate light-dark, plane-void pattern plays out in a rich concatenation of materials and subtly constructed details. Complementing the travertine's pockmarked, red-veined creamy yellow is the velvet smooth bluish-gray concrete of the staircase-office towers, inset with ranges of silvery weathered teak slats. The more we explore and reflect on where we've been and are now, the more complex these distinctive, half-inside, half-outside tower blocks become.

What has Kahn done and why? In contrast to the approach sequence, where he insistently deflected our attention *away* from the buildings and back onto nature's beauty, now Kahn manages forms, surfaces, and materials to emphatically keep our attention trained on these human-scaled buildings, which now seem so at odds with the

A sudden attentional shift, from nature to culture: standing in the Salk Institute's central plaza, looking at the staircase-office blocks

wildness of this natural site. The channel fountain measures just wide enough to comfortably fit a foot. The staircases with their low risers match the step by step of the human body in leisurely motion. Banishing the long internal corridors that are the bane of large institutional and residential buildings everywhere, disaggregating staircases from pathways, Kahn graciously offers up human-scaled sequences and spatial eddies, inviting us to linger and explore. Blackboards on landings signal us to pick up chalk and draw. The private offices are scaled to the size of a standing person; they envelop us in oak paneling that bespeaks domesticity. The carefully laid teak slats on the offices' exteriors betray the art of human making, and the simple, repetitive rhythm of the slats' linear arrangement accentuates the irregular whorl of the wood's grain. All these textural impressions activate and enliven our sense of touch.

Kahn models even the exposed concrete—a notoriously unloved building material—into a luxuriously smooth surface, and the means

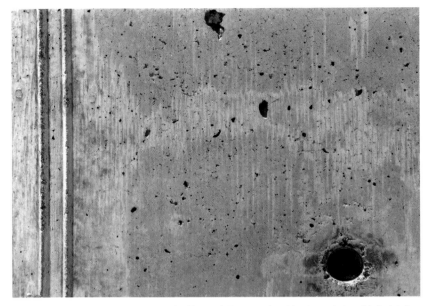

"A building is a struggle, not a miracle. The architecture should acknowledge this": detail of concrete V joints, Salk Institute

of its making offered him the possibility of furthering our identification with the building. Detailing the poured-concrete panels, he told construction workers to sculpt the aggregate that oozed out from between the wooden formwork panels into raised V joints that mark the trace of the construction of the building.

It's a tiny move, like barely visible pencil marks left behind on a finished painting. But just as Neutra describes, it evokes precisely our mental simulation of how that handmade object came to be. Such small details produce large experiential effects precisely because surface-based cues can ensnare us into an intersensory, bodily engagement with the building, and take advantage of our proclivity to mentally simulate a process as a means of understanding an object. "A building is a struggle, not a miracle," Kahn once sighed. "The architecture should acknowledge this."

The smoothly polished concrete. The crafted construction details. The hewn wooden slats. The travertine pavers meticulously laid to expose shadow joints. The precisely dimensioned proportions conveying the building's weight and termination. Carefully executed, all these surface-based cues and indicators of the interplay of materials, gravity, and light enmesh us and the people we see and imagine here, in this and only this place, in a richly constructed moment that engages so many of our senses. Kahn composed the overall forms of the Salk Institute to establish its and our connection to nature, the world, and the subject of biological research. Then he designed the complex's materials and surfaces to pull the thread of nature into and through a world suffused with the grace and presence of humanity.

A Humanistic Bureaucracy: Alvar Aalto's Synthesis of Nature and Culture in the National Pensions Institute

Abiding by the tenets of experiential design can happen anywhere, not just in places as stunningly distinctive as La Jolla's dramatic coastline: it can improve any kind of project and can be employed to create different kinds of effects. Alvar Aalto, a contemporary of Kahn's, is the author of many such projects, mainly in Europe, and some of his most famous buildings, including the celebrated Säynätsalo Town Hall and the Church of the Three Crosses in Imatra, both in his native Finland, continue to influence contemporary designers. Aalto's National Pensions Institute (NPI) in Helsinki, the headquarters of Finland's social security administration, is neither his best-known nor his most dramatic project. Yet in its understated way, the NPI is an arcadia of earthbound light and serenity, illustrating the transformative power of experientially informed design, whereby even the ostensibly banal can be made into a special place.

Occupying a block-sized site in a residential neighborhood north of

National Pensions Institute (Alvar Aalto), Helsinki, Finland: entrance is to left of the front block

Suburban office building, United States

downtown Helsinki, the NPI, at first glance, smacks of a thousand ribbon-windowed, developer-built, suburban office complexes in the United States. But pictures suggest a misleading banality. Experiencing the NPI in person reveals that something as ordinary as a governmental office building, which could have been a body-and-soul-deadening workplace, can be a human-centered oasis that stimulates the imagination and enhances well-being.

Disaggregated volumes mitigate scale: side elevation, National Pensions Institute

To balance the neighborhood's residential scale with a pensions office large enough to serve the entire Finnish population, which, upon opening, employed eight hundred people and includes offices, storage, meeting halls, a library, and a cafeteria, Aalto adopted a deliberately non-iconic approach. Linked rectangular prisms hug the perimeter of the sloping triangular site, and these disaggregated large volumes mitigate the building's large scale. Each prism differs in some way from

Red rock, red brick: National Pensions Institute (detail, near entrance)

the ones adjoining it, offering some indication of the functions it contains: a tall, vertical, tile-faced tower houses the elevator bank and staircases, while blank, brick-faced blocks hold the storage spaces.

Finland, one of the world's most sparsely populated countries, is also one of its northernmost. Nature looms large. Aalto integrates the enormous NPI into its red-rocked, sloping site not only by parsing its functions into seemingly discrete volumes but also by pairing the site's exposed rock with unevenly shaped bricks in deep reds and blackish browns. An earthy, textural palette of materials clads most of the building's prisms; complementing these are partially dressed green-gray granite blocks, edged with copper flashing that weathered long ago to an acid green. The NPI outsizes its neighbors without overpowering them; it seems of a piece with the land on which it sits. The substantial entrance block, approached from the north, opens into the large multistory main meeting hall, where people confer with state employees about their pension accounts. Smaller meeting rooms and offices occupy adjacent blocks. At the northeast end of the site, the NPI's exterior facades create a mostly enclosed courtyard composed of two linked, beautifully planted outdoor garden "rooms" accessible from both an adjacent public park and the agency's cafeteria.

Deft planning, a deceptively modest formal composition, and texturally evocative materials knit the NPI into the craggy topography of its

site. Its highly textured, varie-
gated natural materials elicit
the response of all our senses,
while its easily apprehensible
but varied forms excite our curiosity. Inside the building and
in the main public spaces, Aalto
continues these themes and lay-
ers into them salient embodied
schemas and metaphors that
reinforce the project's themes: a
gentle democracy nurtures citizens modest enough to coexist
with the natural world.

Courtyard with garden, National Pensions Institute

As we approach the main,
north facade, a large brick-clad
rectangular prism is immediately, nonconsciously apprehensible: a geon.
But when we search for the entrance, what at first appeared as a symmetrical volume of contained space reveals itself to be neither fully enclosed
nor compositionally symmetrical. Outdoor terraces at the roof line and
the entry level break the regularity of the box. Entering the building requires that we ascend a staircase located off-axis from the symmetrical
facade's midpoint, slid between a dramatic rock outcropping and a covered terrace: deliberately, Aalto makes us walk a path between nature
and culture. It's an entry sequence that illustrates how even this large
governmental institution abides by nature's dictates.

Inside, Aalto integrates the elements and features of the natural
world literally, schematically, and metaphorically. Daylight reaches even
difficult-to-access places, sculpting and eliciting different atmospheres
from space to space. In the brightly lit main meeting hall, the triple
height ceiling zigzags up and down in a double range of steeply slop-

Double-tier skylights, National Pensions Institute

Baton tiles (interior), National Pensions Institute

ing windows stretching upward, as if they were glassy mountain peaks reaching for the light of the sky. Between the two sets of skylights, hanging cylindrical light canisters punctuate the angular rhythm and, on dark days, boost illumination into the large room. In the NPI's two-level library, elsewhere in the building, circular light wells sculpted into the deep ceiling cast more evenly filtered natural light, suited for reading: as in his Town Library in Viipuri (now Vyborg, Russia), Aalto had calculated the depth of these evenly spaced light wells to catch the Finnish sun's wan rays even in midwinter.

In any very large, multistory building, creating dynamic internal corridors and illuminating them properly presents one of design's thornier problems. Aalto widened the NPI's internal corridors and whenever possible threaded them through places in the building where they could catch the rays of daylight from exterior windows. Like Kahn, and as he himself did in most of his other projects, Aalto also lavishly attended to the NPI's interior surfaces. Corridor walls are clad with long, thin shiny white-and-blue-enameled tiles, flat ranges punctuated with linear ranges of curved, protruding "baton tiles." These linear elements bulk up the building's corridors and pull us along our path, and as they do so, they facilitate wayfinding. And most of all, the baton tiles invite perception as action, practically pulling us toward them so that we might run our hand along them as we walk, feeling their coolness, delighting in how the bump, bump, bump on our hand registers our progress through space.

Inside and out, with these and other design moves and material details, Aalto calls upon our embodied experience of nature in what seems, at first blush, a most ordinary building. As the comparison with the ostensibly similar (in reality witheringly different) suburban office block shows, harmonizing a built environment with our human embodiment in the natural world takes more than the determination to do so. It takes an awareness of human perceptual subtleties, along with the creativity to appropriately accommodate them.

Situating Constructed Environments in Nature in the Twenty-First Century

Today our built environments must rise to the interrelated challenges of hyper-urbanization, globalization, and climate change, making the imperative to take our embodied experience of the natural world into account that much more pressing. But the task need not be any more difficult than it was in 1956, when Aalto completed the NPI, nor more expensive. Consider, for example, Diébédo Francis Kéré's Primary School in Gando, a remote village in southeast Burkina Faso that to this day contains no running water or electricity. Kéré was well

Accommodating climates and building cultures: Primary School (Diébédo Francis Kéré), Gando, Burkina Faso

familiar with the region's climate and customs, having been born to the village chief of Gando, where he lived until he was seven years old. Beginning with the region's vernacular traditions, Kéré selected common materials, mud and corrugated metal. But then he modified existing construction practices to fabricate a clay/mud hybrid brick that proved extremely durable while providing superior thermal insulation and he trained local residents to make them, effectively jump-starting an industry in one of the world's most impoverished areas. He also decided to abide by the common practice of using corrugated metal for the school's roof, but made a simple but dramatic improvement on local construction traditions by raising the roof above the supporting walls. This provided superior ventilation in Gando's searing hot climate and created large overhangs that shade and protect the surfaces of the exterior walls. The result of this budget design? An affordable, elegant learning environment, deeply situated in nature and connected to place.

Built environments that situate us in our bodies and the natural world can happen not only in marquee buildings or in the regulatory freedom of the developing world. They can happen almost anywhere. This includes even the most challenging building typologies, such as high-rise developments. Some people, including some theorists of architectural experience such as Christopher Alexander, continue to insist that tall buildings are intrinsically antithetical to humane domiciles. But that position is resoundingly belied in the human-scale, nature-rich residential and mixed-use projects in cities by Stefano Boeri in Milan, Singapore-based WOHA Architects, and Cambridge, Massachusetts–based Safdie Architects. In Boeri's "Vertical Forest" in Milan, a green project, the equivalent of 500,000 square feet of single-family dwellings is packed into two tall residential towers, one 360 feet and the other 250 feet tall. Over 900 trees and many thousands of plants are arranged, differing from facade to facade according to its orientation to the sun, creating green-filled homes for its residents while reducing carbon emissions, cleaning the city air, and promoting a biodiverse microclimate right smack in a dense area of downtown Milan.

In hyperdense Singapore, WOHA also demonstrates the possibilities of sustainability and design excellence with their two tall residential buildings, Newton Suites and 1 Moulmein Rise. Both largely eschew artificial climate control, relying instead on design to intensify natural ventilation and cooling. The towers are oriented to capture prevailing breezes, and they use modular window-shading devices, balconies, and green walls for cooling and ventilation. These climate-control elements simultaneously establish a body-centered scale and create, from the simple visual language of repetition and variation, an appealingly abstract pattern. In Newton Suites, closely spaced, horizontal shelves of black perforated metal cool the building's surfaces and reduce energy costs. These alternate with two differently patterned vertical ranges of U-shaped concrete balconies, one in a simple repeating pattern, the

High-rise nature: Vertical Forest (Stefano Boeri), Milan, Italy

other arranged in a complex A:B:B:B:A rhythm. All this is surrounded by lush planting and green walls that run the entire vertical height of the facade. Both buildings, when viewed from the surrounding neighborhoods, strike a dignified profile. Moshe Safdie began exploring how to integrate nature into dense urban areas decades ago, in both extensive writings and in his famous Habitat dwellings, of which one version was constructed in Montreal for Expo 67. Recently he began a series of projects updating the Habitat concept; one is his middle-class Golden Dream Bay development in Qinhuangdao, China, in which fifteen-story terraced buildings are stacked atop another at right angles, creating twenty-story-tall framed openings. This open-web, terraced arrangement directs prevailing breezes into and through the apartment units, and maintains views for city dwellers of the Bohai Sea. Just these few

Tall buildings in warm climates: Newton Suites (WOHA Architects), Singapore

Qinhuangdao Habitat (Moshe Safdie), China

examples demonstrate the ways that contemporary designers can integrate our situatedness in our bodies and in the natural world into even today's complex, ever-urbanizing, and global world.

Human experience, including its nonconscious and conscious cognitions, is situated in three dimensions. So far we have discussed two, the human body and the natural world, each of which is a product of evolution. The third dimension, the social world, is less tethered to the dictates of our biological evolution in physical bodies inhabiting a physical world. Humans are also decidedly *social* beings. The individual and social worlds that we inherit and create are strongly influenced by the *places* where our engagements and interactions transpire. Places situate us as individuals among others, and places help us become and sustain ourselves as members of the many overlapping social groups through which we live our lives.

CHAPTER 5

People Embedded in Social Worlds

Every single one of us is a little civilization built on the ruins of any number of preceding civilizations.

—MARILYNNE ROBINSON, *Gilead0*

So strictly regulated and demarcated was life here that it could be understood both geometrically and biologically. It was hard to believe that this could be related to the teeming, wild, and chaotic conditions of other species, such as the excessive agglomerations of tadpoles or fish spawn or insect eggs where life seemed to swarm up from an inexhaustible well. But it was. . . . In the same way that the heart does not care which life it beats for, the city does not care who fulfills its various functions. When everyone who moves around the city today is dead, in a hundred and fifty years, say, the sound of people's comings and goings, following the same old patterns, will still ring out.
The only new thing will be the faces of those who perform those functions, although not that new because they will resemble us.

—KARL OVE KNAUSGAARD, *My Struggle* (Book I)

magine this slightly sci-fi scenario. It's a temperate spring day, and we have a couple of hours to kill. In this world, intercontinental travel happens in an instant, and we decide to go for an exploratory walk. Three locations present themselves: one in Paris's Latin Quarter, the next in Jerusalem's Old City, and the third, in downtown Seoul. You remain the same person, but nations, cultures, and streetscapes differ, and so too will your conduct and cognitions.

In the cool green gardens surrounding Paris's Luxembourg Palace, strolling tempers the pace of our thoughts. Muscles soften, expiration slows. The Luxembourg Gardens' design releases us from the burden of making decisions, as pebble-strewn paths direct our crunching footsteps here, then there, presenting glances of tightly manicured shrubs, sweeping our gaze toward the butterflies alighting on flowers. In the distance we catch a glimpse of the nearby Panthéon, a large eighteenth-century neoclassical sanctuary that was rededicated by French revolutionaries to the memory of their heroes. Among the luminaries interred there are Voltaire, Jean-Jacques Rousseau, and Victor Hugo.

Deciding to visit, we circumambulate the busy roundabout, resuming our leisurely pace on Rue Soufflot, named after the Panthéon's architect, Jacques-Germain Soufflot. This street traces an axis directly to the facade of his hilltop masterpiece, framing its view. Little distracts us from our destination as we ascend the inclining street. The ground-floor spaces of the stone-faced neoclassical apartments contain the occasional sparsely populated sidewalk café, a few sleepy restaurants, a pharmacy, and a bank. The urban soundscape quiets as we ascend the

Below: Panthéon (Jacques-Germain Soufflot), Paris, France
Opposite: Interior, Panthéon

street's uphill path, passing by the law school of the University of Paris. As we arrive, the Panthéon dominates its quiet plaza, its dome towering above, rising, at its peak, 136 feet taller than the towers of Amiens Cathedral.

Inside, a resplendent colonnade of paired white Corinthian columns soars into airy vaults sprung from pendentives; it's a composition that can't but draw our eyes to the heavens. The Panthéon, like Amiens Cathedral—and for some of the same reasons—immerses us in that all-too-rare moment. Time stops here. These incised, lofty spaces are devoid of pews, altar, choir screens, or pulpit. In the transept, Foucault's pendulum swings an arc that traces the minutes and hours of the earth's rotation. If Amiens Cathedral inspires reveries of an otherworldly god, the Panthéon's message seems to be that here science, not deities, reigns.

Surrounding us is a restorative landscape wholly distinct—due to its white precision and restrained serenity, drenched in an unearthly white light—from the restorative greenscape of the Luxembourg Gardens, where we began. We drink in the Panthéon's silent dignity.

Now imagine that, instead, we spend the very same afternoon in Jerusalem, on a walk of similar duration. We begin at the Jaffa Gate in the mainly pedestrian Old City. A uniformed klatch of Israeli soldiers—barely adults—smoke cigarettes, their automatic weapons serving as armrests, unheeded at their sides. Beyond the fortified entrance, we nearly stumble right into a group of backpack-laden teens, clearly tourists, huddling around a tiny alcove sparkling with Christian trinkets. The proprietor shuttles between languages as he earnestly communicates directions to the nearest hostel. Here we stand, at the holy crossroads of a succession of civilizations, swimming in sanctity, simmering with political conflict. Five, perhaps more, cobbled alleyways offer up their shadowed paths, intimating any manner of fulsome experiences. We decide to forgo a visit to the iconic Dome of the Rock, the Church

Historic photo of Jaffa Gate, Old City, Jerusalem

A riot of visual and auditory stimuli: souk, Old City, Jerusalem

of the Holy Sepulchre, or the Western Wall, approaching instead the nearby souk, or Arab marketplace. A cavalcade of shopping opportunities flicker through our imaginations.

Makeshift stalls and linoleum-floored stores disgorge cheap jewelry, brightly colored scarves, and muted tapestries matching and unmatching the saturated hues of powdered spices, piled high. Merchants beckon. Come, come here—you don't want dried fruit? A bracelet for your daughter? A trivet for your pot? It seems that not one of the merchants in these corrugated-metal-covered corridors would survive the day deprived of our company and patronage. After fifteen or twenty minutes of declining countless such invitations, we change plans. Too much. Lefts and rights through shadowed alleyways expel us, finally, into the hush of the beautifully maintained Jewish Quarter, lined with yellow blocks of creamy Jerusalem stone. Carefully we pace its uneven surfaces and wander its stone-dressed passages, passing under carefully groined vaults supporting some family's extra room, peering into plant-filled courtyards. Suddenly, we happen into the sky-roofed, stone-clad Hurva Square, which surrounds and presents the neighborhood's largest synagogue. Tourists sit on parapets, smoking cigarettes. Children jump the benches, the tassels of their zizith flying behind them, while clusters of black-clad men and women, presumably their parents, look on, chatting, passing the Shabbat holiday with friends.

Cities as Social Worlds

Accidents of history and acts of commission both have made these distinctive streetscapes and neighborhoods in Paris and in Jerusalem exact their pull on what we think, feel, do, and decide. In one way the point couldn't be more obvious: we wouldn't go to Paris to see Jerusalem. But the pervasive, decisive influence of place on what we think, feel, decide, and do can be difficult to recognize. During these few hours in both cities, for example, we are alone, but only in Paris does solitude define our experience—because no one talks to or approaches us, nor we them. And in Paris, unlike in Jerusalem, the combination of the city's muted sounds with the cityscape's clear organization—with paths, edges, nodes, and landmarks all in evidence—liberates us from clattering distractions, freeing our minds to meditate on larger themes. Placards inform us that the awe-inspiring Panthéon was completed in 1744 by Louis XV as a church in honor of the patron saint of Paris, Sainte-Geneviève. During the National Assembly in 1791, the building was redesignated a public mausoleum commemorating the revolution's heroes, celebrating democratic freedom and the secular state. This knowledge of the Panthéon's origins turns our cognitions revolving around politics, religion, and history: monarchies decapitated by revolutions, in turn eviscerated by dictatorships, overthrown by monarchies . . . and eventually, democracy.

Even though in Jerusalem's densely packed Old City, political turmoil openly simmers, the Jaffa Gate's tightly packed everyday quality discourages, indeed precludes such focused contemplation. College-aged soldiers joke with one another and nod to you. Each shop seems crammed with more stuff than the next. Stone paving requires us to move deliberately, watching our step. Converging here are people claiming all manner of allegiance: ethnic, geographic, political, religious—Palestinians, Jewish Israelis (secular and orthodox, armed and unarmed), Palestinian Israelis, Christian pilgrims, Armenian priests, and polyglot tourists from Africa, Asia, Europe, America—all clad in the visual argot of their cultural affiliations, and moving in an unending dance of yeses and noes, of staking out and ceding ground, of selling and buying, of marveling and feigning delight or apathy or disgust. It's a surfeit of spectacle and opportunity.

Perhaps, instead of Paris or Jerusalem, we've chosen to walk in downtown Seoul's Insa-dong neighborhood. Most of Seoul, with its high-rise

Insa-dong District, Seoul, South Korea

steel-and-glass buildings, could be Bangkok, Dallas, or any major modern city. Not Insa-dong. Small commercial blocks and diminutive townhouses are crammed against one another. On some parts of the street, the distinctiveness of individual buildings is drowned out by the colorfully illuminated signs hanging off the facades, yellow and orange and blue and green vertical banners advertising computer repair shops, clothing boutiques, and antiques. Insa-dong is a thoroughfare that Seoul's city planners have curated to retain some of the cluttered feel of an older Asian street. It is closed to automobile traffic on weekends. Seoul's residents treat it as their own pedestrian mall every day of the week, so we follow their lead, walking to and fro, window-shopping, enjoying the street's illuminated signs and the buildings' carved doorways and windows, projections and recessions.

A newer retail building, larger than its neighbors, anchors the street corner in a composition of slatted wood planes, gridded concrete prisms, bands of windows, and textured, projecting cubes. A few

Ssamziegil (Moongyu Choi + Ga.A), Insa-dong District, Seoul

blocks later, we come upon a concatenated retail complex: its ground floor displays a grass-covered range of small shops, and rising above is a textured brick prism with the unpronounceable *Ssamziegil* written in Latin characters above its reflected, double-inverted Y logo. Moongyu Choi, the architect, designed the midblock entrance so that it opens into a courtyard. A wide bank of amphitheater-like stairs invites us to ascend.

Unlike shopping at and standing among the cacophonous market stalls of the Old City, here, we're intrigued. Steep steps lead to a gently inclining ramp, which wraps around an open interior courtyard. Minuscule boutiques and serene restaurants line our path. Each time we enter an artsy shop, its proprietor smiles demurely; such graciousness, in the absence of other shoppers and in such close bodily proximity, makes it seem impolite to ignore. So we strike up a conversation. One shopkeeper, we learn, is earning a degree in art and fashion from Seoul Women's University; shyly she acknowledges that she designed every

Above: Shops edge a gently inclined ramp, Ssamziegil

Below: Textures, patterns, and materials in hanok facades, Bukchon District, Seoul

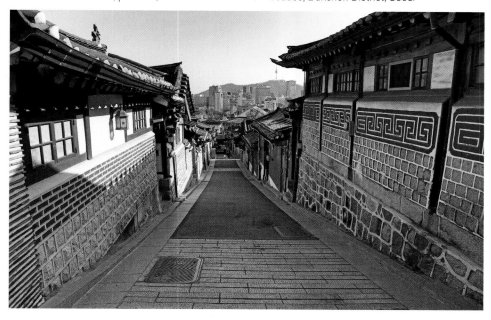

Opposite: Entrance, Ssamziegil

article of clothing on the racks. Partly to support a nascent artist, partly to kill time, and partly because we are simply charmed by it all, we purchase a gift for a friend before wandering out of the Ssamziegil complex. Within fifteen minutes, we've reached the nearby historic district of Bukchon, where lavishly restored *hanoks*—the traditional Korean courtyard dwelling—line narrow streets.

Each hanok facade has been lovingly restored into a collage of textured geometric patterns in bands of color, punctuated with solids and voids. Blazing white mortar holds mottled masonry blocks and enameled bricks in deep red and charcoal gray in place. Above, wood-framed squares of windows are recessed below projecting joists, with doubled ranges of terra-cotta tiles terminating the compositions. An elegant concert of repetition and variation ushers us in toward the entrance to one of Bukchon's many teahouses, where we find wood-paneled rooms surrounding a beautifully proportioned courtyard. Task lights and bookshelves line the walls, and large tables and cushioned benches fill the room: the very scene makes us determine to take the afternoon off. We settle among the other urban professionals with a pot of tea, anticipating a serene afternoon of reading and correspondence.

Social Worlds Are Action Settings

Three cities. Same season. Same time of day. Same activity, even: a stroll in a pedestrian-friendly historic neighborhood. Same person! And yet how different our afternoons have been. This includes both our internal cognitions and experiences as well as the way we've related to the people around us. Even when doing the same thing—shopping—it's different. In Jerusalem's Old City, we constantly shift our heads and torsos this way, then that, seeking out the stranger whose package we inadvertently sideswiped, or declining an invitation to survey someone else's wares. Maintaining silence or the distance of personal body

Teahouse in a renovated hanok, Bukchon

space would be impossible. Everyone's talking to, jostling, eyeing, and elbowing one another. In Seoul's Insa-dong, shopping in the absence of other customers, we savor the store's quietude and the reticence of its proprietor, which inspire us to strike up a conversation. And Paris's Latin Quarter, compared with the Old City and Insa-dong, is empty. The only conversation we hear is the one inside our heads. In the Pan-théon we feel a kind of connection to others too, but abstractly—our sense of commonality is to humanity in general rather than to specific people. That emerges from our experience of the building's grandeur: awe, remember, promotes in us other-directed, prosocial feelings, our awareness of people's shared humanity.

Just being in an environment that is arranged in a patterned way— the Latin Quarter's wide sidewalks, the Old City's irregular streets and passages, Seoul's ramped mall and even hanok village—prompts us to think, feel, and act differently. That's true of every place we encounter. A formal garden, a vaulted sanctuary, a souk, a courtyard mall laced with high-end boutiques, a historic district; each encourages us to engage in specific activities and to think certain thoughts. And each discourages or all but precludes other activities or thoughts. Indeed, as Karl Ove Knausgaard captures in the epigraph to this chapter, the configuration of our built environments so powerfully predisposes us to act, feel, and interact with others in specific ways that, were we to replace all the people currently in the Luxembourg Gardens, the Old City's souk, or Seoul's Bukchon with a different set of people, the general patterns of their actions and interactions would be the same.

Buildings and interiors and streetscapes and landscapes are all *action settings*, places shaping what people do and think and how they engage with one another. Every action setting is composed of what we have been calling affordances, meaning spaces and objects that afford certain actions. (A living room, for example, is an action setting. The couch is an affordance in that it *affords*—it both suggests and facilitates the action of sitting.) Action settings, such as neighborhoods or clothing stores, contain patterned arrangements of objects and spaces. Those arrangements provide us with informational cues that are critical to our lives among other people, influencing us to act in patterned and socially normative ways.

The notion of action settings draws on the work of Roger Barker, one of the forgotten founders of environmental psychology. In the 1950s, Barker launched a full-scale critique of behavioral psychology, arguing that because behaviorists (such as B. F. Skinner) confined their psychological research to the laboratory, they inadvertently neglected an entire dimension of human experience that profoundly shapes human behavior: the environment. With colleagues at the University of Kansas, Barker set up the Midwest Psychological Field Station in 1947, and for nearly thirty years conducted studies of people's behavior in situ. In one study, researchers outfitted with pens and notebooks followed children from morning until night, from home to homeroom to class to the cafeteria to playground to classroom to soda shop, and then back to home.

Predictably, the children's conduct changed during the course of the day. Less predictable was Barker's finding that one factor overwhelmingly determined the children's patterns of behavior: location, where they were at a given time, and how that place was configured. How Jessica and Sabrina acted in the classroom predictably differed from how they acted at assembly. What Adam and Aaron did at home predictably differed from their conduct during chess club. This may not be so surprising, really, but here's the field station's other finding: Barker and his colleagues found that they could *better predict a child's conduct at a given time by specifying her environment and its action setting* than they could by delving into her individual, psychological profile. Barker wrote that the "variability in behavior of different children in the same setting at a particular time was smaller than the variability in behavior of the same child across his or her entire day." And just as Jessica's and Adam's conduct varied depending upon where they were situated, so also, it is safe to say, did their conscious thoughts and decisions, and their nonconscious cognitions and emotions. In unraveling the mystery of human conduct, action settings had been a critical hidden variable.

In our experiences of the built environment, action settings constitute our social worlds. The conduct studied by the Midwest Psychological Field Station was situationally patterned because the children were doing what a person nearly always does in an action setting: following the very small set of stated rules and the much larger constellation of unstated norms. Both norms and rules are established in and perpetuated by institutions in how they and the objects in them are designed, selected, and arranged. When we enter a built environment, we immediately survey and apprehend its character, figuring out, usually with a glance, what we can and should do, discerning all the norms that govern people's conduct there.

By taking psychological research out of the laboratory and into the world, Barker and his colleagues resoundingly established that if we want to understand not only people's collective actions but also their internal lives, a deep analysis of their action settings—our human habitats—must ensue. Our afternoons in Paris, Jerusalem, and Seoul illustrate the many ways that action settings, conventionally called places, inflect what we think, feel, and do, including what we do with other people. Whether we amble the expansive sidewalks of an elite metropolis in Europe, shoulder our way through throngs of tourists in an ancient Middle Eastern city, or dart from place to place in a human-scaled part of one of Asia's supercharged megacities, our thoughts and actions and choices and social engagements will differ. Most important, they will differ in *patterned* ways.

So the Midwest Psychological Field Station and Barker's findings lay bare the fundamentally social nature of the built environment: inhabiting permanent structures in congregations of settlements is part of what constitutes our humanity. Because our children take such a very long time to develop into adulthood and require so much care and training to properly mature (parents know!), and because our inordinately large brains require a mix of high-octane nutrients that comes only with cooked food, humans, throughout the life cycle, came to rely on one another. Children need to be watched, fires needed to be built and tended, everyone needs protection while they sleep. To be a human being means to be a *social* being. People so fundamentally crave the presence and companionship of others that if society consigns a person—any person, from a psychologically healthy college student to an incarcerated felon—to solitary confinement, she will suffer symptoms of psychosis after just a few days, and the longer the isolation, the more acute the derangement. Evolution amply rewards our innate sociability: the denser our social relationships, the healthier and longer we are likely to live.

Built settlements enabled our ancestors to live in stable groups, and these groups grew in size and evolved in complexity, as did the activities in which humans engaged. Growing economies encouraged people to differentiate vocationally, so that one person could become a dressmaker, another a baker, another a soldier. Heterogeneous economic activity, combined with evolving social and political institutions, redounded in ever more complex and ever more differentiated built environments. So while the earliest human settlements featured only homes, sacred places, and marketplaces, in time followed buildings for manufacturing, gathering places, schools. And sports arenas, civic places, auditoriums, courts of justice, and on and on . . . resulting in cities, which grew and changed, and continue to grow and change today. The action settings of these places make manifest the religious, political, economic, and cultural institutions and traditions they house. In doing so, they advance and perpetuate prosocial conduct and norms. In this way, people's sense of membership in and loyalty to social groups is writ in stone, so to speak: the built environment itself facilitates and sustains social life, and helps to perpetuate the social order.

Home as Refuge: Place Attachments in Action Settings

When people lay claim to a piece of land, constructing buildings, organizing and shaping its voids into action settings, it is no longer just an abstraction, a geographic point on a map. What was once just territory becomes a *place*, which means that it is imbued with social meaning. People with stronger attachments to places enjoy an enhanced sense of well-being, stronger community ties, and a greater ability to transcend their own self-interests and conceptions and see others' points of view. Cities are collections of places, with a society's institutions, with their action settings and patterns of human interactions, distributed across space.

How do each of us come to experience a given place as the setting for this and this activity, and not that or that? How does the built environment *situate people in place,* socially? To answer this, it makes sense to begin with the one constructed place—home—where each of us effortlessly connects the dots among four things: our internal, private experience; a social grouping; a physical construct; and a set of patterned activities. If our experience of the built environment is situated in our embodied selves, and our embodied selves are situated in the physical principles and ecosystems of the natural world, then humans are—by definition—situated also in the social world. In that human social world, the home anchors society's smallest, most foundational institution: the family. A home protects us from inclement weather and other unwanted intruders, both animate (people, animals) and inanimate (noise, unwanted light). And unless a person or a family meets inestimable misfortune—which more than 550,000 homeless people in the United States currently have—its members reside, at least some of the time, in a home. That's 320 million plus Americans.

A home offers much more than shelter. It centers us and our family members, literally and psychologically: after a long day or a long journey, it is to a home that we eventually return. Rebecca Solnit writes that home is the place where you are "the point of intersection of all the lines drawn through all the stars." And if our evolutionary heritage mandates that we "prospect" where we can, it also ensures that we take safe refuge—if not in reality, then certainly in our imaginations—at home. As action settings, homes afford us a wide-ranging but restricted set of activities and states of mind: refuge from our prospecting, orderliness, freedom, sociability, and privacy. We enjoy more autonomy, more control over our surroundings at home than we do anywhere else. We are free to shape and decorate the environment, to establish routines and break them, to be ourselves, alone or with others. As we've seen, that sense of autonomy is fundamental to our health and well-being. When a home fails to provide any or all of these qualities, its inhabitants, especially if they are children, suffer. There are acute and lasting developmental, cognitive, and psychological consequences.

A home, then, is many things: a place on the land, a building, a

Home schema: Tokyo Apartments (Sou Fujimoto), Tokyo, Japan

psychological concept, and a container serving a small and defined social group. Homes facilitate certain experiences and impede others. Because of the density of these combined experiences over time, people tend to develop profound attachment (usually positive, but sometimes negative) to these specific places that serve as their homes, and more generally to other, related places that they frequent, and with which they come to feel familiar. More likely than not, this phenomenon, which psychologists and geographers call "place attachment"—the affective bonds we develop with the places and spaces of our worlds—constitutes an elemental need, analogous to an animal laying claim to a given territory. The density of our contact with a place and the intensity of emotions we associate with it determine the quality of our attachment.

Place attachment is central to identity formation. As children, we spend most of our time at home, and by the time we reach just ten months of age, we easily differentiate familiar from unfamiliar places. The stories and narratives we develop in these years, at home, are the ones that we write and rewrite all our lives; they are the stories of what we have seen and done and been, when we were in the places and spaces and buildings of the world.

Residential architecture's panoramic history illustrates that a wide range of styles, materials, and spatial arrangements can embody the home's physical, social, and psychological qualities. Even so, the home is a classic example of one kind of action setting; like all action settings, some of its elemental features are constant. The four walls plus roof that children around the world draw as "home" suggest that, if nothing else, the human schema of home as an action setting involves a minimal guarantee of shelter: a home *contains* and *protects* us. Contemporary architects playfully reinvestigate the home schema in various ways: Sou Fujimoto stacks such volumes for the tiny Tokyo Apartments in Japan, and Herzog & de Meuron use a similar motif for its high-end home fur-

Home schema: Vitra showroom (Herzog & de Meuron), Weil-am-Rhein, Germany

nishing store, the Vitra Retail Building in Germany. People's schemas of domestic interiors vary much more widely, of course, but very often contain larger, open common areas to promote social intercourse, and smaller, closed sleeping areas.

The strength of our affiliation with a place and the institution it houses and represents—even whether or not we correctly understand its character—can be fostered or encouraged by design—and through design it can be discouraged and even destroyed. Take the experience of a diplomat, Rob McDowell, living in Vancouver, recounted by Charles Montgomery in his book *The Happy City*. McDowell bought a condominium on the twenty-ninth floor of a high-rise residential complex that also contained larger units in townhouses on the ground floor. McDowell was drawn to the flat because it captured views of the city's stupendous ocean and mountain landscapes. But he found, after living there for nine months, that his social life was not to his satisfaction; walking from his condominium to the shared space of the corridor and into the elevator offered him few opportunities to get to know his neighbors. So McDowell relocated to one of the townhouses in the same complex. There, each home's front door led to a porch overlooking the common garden.

In the tower, spaces were either private (one's own apartment) or public (corridors and lobby). Such an arrangement meant that to engage socially with neighbors, McDowell risked intruding on someone when she might just want to escape to her private refuge, thereby violating social norms. By contrast, the design and landscape of the townhouses' exteriors carved out the kind of semiprivate area that facilitated a more productive kind of social engagement. His porch provided a space that was shallow enough to encourage McDowell to strike up conversations with strangers, because both he and they knew they could beat an easy retreat without having to explain. Living in the same residential complex, but in a unit with a slightly different exterior design, literally changed McDowell's life. Within a decade, he was counting half of the twenty-two fellow residents of the townhouses among his closest friends.

Action Settings in the Social World

As children constructing a "home" schema, we associate the particular kind of place that a home is with three categories of things. First is the patterned set of activities that the home promotes and accommodates, such as sleeping, eating, socializing with family members, and relaxing.

Second is the set of social norms guiding conduct in that specific kind of place. And third is our own individual experience. Through this constellation of associations, home becomes the paradigmatic action setting through which other action settings—other places—are defined. Our home schema informs us that dwelling in a home among family differs from learning at school, an action setting that promotes social congress and disciplined focus. And these differ from the workplace, with its attendant pressure to perform or sense of accomplishment, and all these differ from the taut energy of hip Insa-dong or our serene awe in the ethereal Panthéon. Just as we developed a deep psychic bond to our childhood home, as we grow and explore and experience the world, each of us builds an individualized storehouse of place attachments and affiliations, depending upon the places and action settings that life casts our way.

What holds for our attachment to homes pertains to our attachments to society's many other institutions, urban areas, and landscapes as well. Three factors guide our decisions about whether or not to engage with that particular place and in what manner: how a place's design facilitates human activities and the correspondence among those activities, the patterned arrangement of objects in spaces, and the associations its forms elicit. Action settings thus comprise the mental and physical frameworks we use to understand and make decisions about how to interact with our environments and with others in them. And all this has profound implications for the health and fabric of a society. Visiting the Panthéon elicits in us thoughts of humanity's commonality. Fleeing the Old City's souk in rejection of its cluttered celebration of materialism separates us also from its merchants. Through design, our bonds with the social group that an action setting implies strengthen or diminish: borrowing a concept from behavioral economics, we could say that action settings "nudge" people toward certain actions and conduct rather than coercing them. They promote what Barker described as "situationally normative" conduct.

Action settings offer multiple uses: National Opera and Ballet (Snøhetta), Oslo, Norway

Most action settings offer multiple opportunities and ways to engage them. The most successful ones contain affordances that obviously support recognizable action patterns, as well as displaying identifiable boundaries and a visually coherent arrangement that articulates the character of the place. None of these need be obvious or hard-edged, as Snøhetta demonstrates in its home for the Norwegian National Opera and Ballet in Oslo. Although the large building was constructed as a nighttime venue, Snøhetta turns the National Opera into a kind of public square by tipping its roof in two directions to make its edges the place where city meets shoreline, turning the exterior into an accessible outdoor arena. Similarly, Bukchon's tearoom can be a place to get lunch, a place for a group of friends to socialize, a place to read or draw in solitude. In some cases, the range of activities that an action setting offers can be at odds with one another. Just as a bonfire affords both warmth and destruction, a cafeteria can afford its users a noisy, bustling school lunch or an orderly class assembly.

Because our cognitions, our embodied minds, and our bodies are situated within and wired to be subject to the suggestions of action settings, the way people experience built environments becomes an unending dance of nonconscious and conscious positionings within the constraints and opportunities offered by us in our social worlds. To be clear, this differs fundamentally from the folk model of cognition's misleading notion that the choices we make as we move through the world are by and large conscious. In reality, people are constantly bombarded by the information we nonconsciously glean from the built environment and its objects, which stimulate us to think unbidden thoughts, feel unbidden emotions, and make (at best) half-conscious choices among socially patterned sets of conduct.

The concept of action settings helps to connect the dots between an individual person's inner, and seemingly private, experience of the built world and people's experience of their environments as social beings with group affiliations. Recall, as we discussed in the introduction, that most of the existing literature on how people experience built environments focuses *either* on people's inner experience of the environment (through studies in environmental psychology or phenomenology) *or* on their social action patterns as they position themselves as members of one or several overlapping groups (through studies in urbanism, sociology, and ecological psychology). The notion of action settings eradicates that artificial division by defining our internal experience of built environments as irradicably social. Take the Bukchon tearoom with its cushioned benches and sliding windows. When a window falls within our peripersonal space, our premotor neurons fire, but the patterned physical arrangement of its interior and the patterned social conduct of its patrons inform us that it would not be appropriate for us to open the window. Conceptualizing the built environment as a living ecology of affordances nested within action settings vanquishes the Cartesian folk model of human cognition once and for all, demonstrating that, from the point of view of how people actually experience their environs, the idea that we make only conscious choices couldn't be more wrong. We use places and the objects in them to conceptualize and choose among the action sequences that become our experience, feeding into the narrative stream of our lives.

The GSD: A Pedagogical Action Setting in Action

To illustrate how action settings and the social worlds they embody inflect our environmental experience, let's take a scenario with which

Gund Hall (John Andrews), main facade, Harvard University, Cambridge, Massachusetts

every reader will be familiar: going to school. The Graduate School of Design (GSD) at Harvard University is unusual in that it serves mainly graduate students who study design. Even so, the GSD serves our purposes splendidly because its architect, John Andrews, designed its building, Gund Hall, with a clear vision of how a design school should operate institutionally, educationally, and socially. Gund Hall happens to be a place I know extremely well. I worked there as a professor for ten years.

Finished in 1972, Gund Hall is the loudest and youngest in an architecturally cacophonous group of buildings anchoring the northwest quadrant of the heart of Harvard's sprawling campus. Across Quincy Street rumbles the High Victorian Gothic Memorial Hall, which houses undergraduate dining facilities and a large lecture hall in a massive, decorated brick and terra-cotta building that walks a tightrope between august and carnivalesque. Across the street from Memorial Hall rise four buildings: a small stone neo-Gothic church; William James Hall, a white, 1960s fifteen-story concrete tower designed by Minoru Yamasaki (the

Gund Hall, rear facade (trays housed in pitched skylit shed at left)

architect of the doomed World Trade Center towers); Adolphus Busch
Hall, a horizontally spreading, Germanic gray stucco pile; and the tidy
august Greek Revival Sparks House, a wooden structure that used to be
painted in thickly saturated daffodil yellow and an equally eye-searing
white. The stately five-story Gund Hall faces Memorial Hall's apse-like
rear end, a classic study in contrasts. Gund Hall recalls in equal parts
Albert Kahn's huge, early twentieth-century automobile production factories in Michigan; the stripped neoclassicism of the 1930s (for example,
the Palazzo dei Congressi in the EUR Complex outside of Rome); and the
reinforced concrete, industrially inspired abstractions of Le Corbusier, as
in his Convent of La Tourette, in southern France.

Most of the hundreds of times that I walked there, I focused mainly
on my own immediate goals, musing about whichever essay or book
I was writing, fretting about the unfinished state of my next lecture,
thinking about an upcoming committee meeting, mentally reviewing
the week's schedule. Only now, as I reflect upon my experiences there
in light of what I have learned from researching and writing this book,
do I realize many of the ways that the design of Gund Hall and this
part of the university's campus shaped my conception of Harvard as a
university, of the GSD as an institution within it, of architectural education, of teaching and its meaning in the context of a highly competitive
professional program, and much else.

The visual cacophony of Gund Hall's urban surroundings surely
helped to prime my impression of the GSD's place in the university as
a large, nearly self-sufficient ship amid an uneasy collection of slightly
wayward boats, each stubbornly navigating its own way. Indeed, the
building manifests in its very architecture the university's institutional
and economic organization, described by the Harvard administration
as "every boat on its own bottom." For me, the heterogeneity of Gund
Hall's urban context encapsulated a mental paradigm of the university
as an archipelago of semiautonomous islands. Within this academic ar-

chipelago, the GSD is often described as an ocean liner—Le Corbusier's beloved, now anachronistic icon of industrial technology. Once you get on, it's tough to disembark before your journey is complete. Its world becomes your world.

On the main facade, Gund Hall's staggered lower three stories recess behind its top two, concrete-encased stories, a dark ribbon of office windows stretched across each. This arrangement creates a sheltered portico supported by tall, slender columns, into which various glass-encased offices and the library protrudes. Approached from the opposite end, Gund Hall resembles a glazed hothouse in exposed concrete and tubular steel, rising in a dramatically steep zigzagging slope that steps from its single-story rear facade up to the main five-story entrance facade facing Memorial Hall. Amid the heterogeneity of its urban surroundings, a building of such large size and unusual composition must house an institution of some distinction. But what institution? What kinds of action settings lie within? From here, the only way to know for sure would be to consult a reliable source. But anyone with even a slight knowledge of architecture and its history would ascertain the compositional tropes elicited by design, communicating much about this institution's nature, functions, and self-conceived identity and place in Western culture. Gund Hall's main facade, resembling classical and neoclassical porticoes from the Parthenon to the United States Supreme Court, exhibits a rhythmic, symmetrical composition built up from discrete identifiable components of offices and larger loft-like spaces: the very regularity of this controlled composition suggests that an important, purposeful, historically aware institution is housed within. Yet in contrast to these historical allusions, Gund Hall's surfaces insist upon their modernity. Steel, large plates of glass, and reinforced concrete are all materials architects adopted in the twentieth century as icons of modernity.

In compositional details, the portico's design ostentatiously departs from its neoclassical precedents as well. Local, smaller-scale asymmetries pull the main facade's apparent symmetry apart. And these asymmetries help deliver information about the action settings housed within. Unlike traditional neoclassical exteriors, in which the exterior composition betrays little of interior functions, Gund Hall's main facade exhibits its interior spaces and gives hint of their probable functions. The large, glazed space on the ground floor must be the library. Smaller and larger offices are upstairs, on display.

Through rapid gist identification, we've noted that the overall composition of the portico's colonnade alludes to classical architecture, while Gund Hall's materials, proportions, local asymmetries, and lack of applied ornament establish its decidedly nonclassical, modern credentials. In other words, we've been primed to conceive of the GSD, the institution housed here, as at once saturated in tradition and determined to reconceptualize architectural traditions for a modern world. Andrews's design perfectly reflects the GSD's self-perceived identity.

At the same time, Gund Hall's materials, construction details, lack of applied ornament, and entry sequence contributes to our impression that GSD emits abundant energy but little warmth: an important, vibrant, creative, and singular place of determined purposefulness, to be sure, but neither relaxed nor particularly welcoming. Measured against the scale of the human body, Gund Hall objectively looms large, and seems even larger still because of its vertical proportions. No doubt about it, we have arrived at a *big is important* place. The cold, hard surfaces (large plates of smoked glass, exposed, smooth gray concrete, and black-enameled metal), the jagged edges of the building's enormous stepping skylight, and the exterior's unusually dramatic diagonal sweep provoke in many people a subtle, perhaps barely perceptible stress response. It certainly did in me.

Gund Hall's grand scale and large size encourage us to see the building and the institution it contains as one and the same; it conveys a sense of tough urgency, communicating that all who enter here are witnesses to or participants in an important, creative mission. In all these ways, the design of Gund Hall exemplifies how a place shapes our cognitions and our emotions in both subtle and obvious ways. The distinctive aesthetic of the GSD's container, Gund Hall, communicates in every manner but words that all who enter here participate in the world called Design.

Often, as I traversed the portico's long axis to reach the GSD's main entrance, I would pass a couple of students chatting. Not much invited me to stop. The transition from the street to the interior foyer prompts a sense less of welcome than mission. Why? Neither inside nor outside the front entrance do we find benches or parapets or tables for sitting or gathering. The GSD's spatially amorphous lobby, with its low ceiling, lack of interesting views, and absence of furniture, indicates that you may be welcome in the public area, but not for long. Exhibitions on the walls invite brief looking, but nowhere are there the kind of spatial eddies that encourage small-group congregation. So GSDers mostly act accordingly. Overwhelmingly, people interpret this entry as a transitional

Foyer, Gund Hall

space: students en route to their studios, professors and administrators en route to their offices or lecture halls.

On the days that I entered the GSD's foyer, I often deliberated among the action settings, and social engagements on offer, wondering whether to make my way directly to my office (where I might capture a few silent moments to myself) or take a sharp left into the glass-encased library (where I could look up a reference for that essay I was writing) or gravitate diagonally toward the noisy force field of the café, which opens into student-crammed loft spaces where the design studios are held. Deliberating meant imagining myself being among others in those places—library, café, office, studios—doing or not doing what is done in each.

In the café we become party to the bustling, active, *content*-filled loft spaces that GSDers call the trays, by which they mean the five open floor plates stretched lengthwise across the building, which house the school's many design studios in a vertically slung "open classroom." The design of these studios extends the open classroom concept to its

The trays, Gund Hall

logical apotheosis; it's an open school. From anywhere in the trays, Gund Hall resembles an enormous skylight-turned-production facility. The trays create linear yards upon yards crammed with students, professors, drawing tables, stools, desktops and laptops and notebooks, locked supply cabinets on wheels. Pinned up on its partitions are untold miles of sketches on drawing paper, photographs, postcards and Post-its, computer printouts of digital designs. Students typically sit at their drawing tables, working by themselves or conferring with at most one other person. Small-group gatherings happen when an instructor makes an announcement or presents material.

As action settings, the design of Gund Hall's studio spaces, or trays, directly shapes both the learning and the social life of the GSD students. The trays do this in ways both good and less than ideal. Since every student's project, in all its various stages, is quite literally within view of the others, the trays virtually ensure that students will come to see the making of a building or a landscape or an urban design as a *process*. Students rework their projects constantly, or so it seems. And because the

Few places to gather: professor instructing students in Gund Hall's trays

trays place all the studios together in one vast open space, with all the other students in view, the place emits a kind of kinetic energy, imparting the sense that everyone participates in a large, vibrant community of colleagues and friendly rivals, with everyone committed to and engaged in the shared vocation of making a better built environment. Andrews's division of the design studios into multitiered spaces also breaks up Gund Hall's objectively large scale: if all the GSD's studios were placed in a loft space on a single level, this place would overwhelm us with its immensity.

But Gund Hall's design suffers from what an economist might call its social externalities—the unintended consequences of Andrews's open school concept. The GSD's competitive atmosphere is legendary. Students work incessantly. Sequential all-nighters are common. Ambition seeps through the trays, and the number of students who want to be, who expect to be, the next Rem Koolhaas or Bjarke Ingels is staggering. Many are quite self-consciously positioning themselves to earn the favor of professors or the school's many visiting designers, architects serving as critics, whom they correctly identify as their easiest tickets to employment after graduation. No matter how many times faculty emphasize (or don't) that design is a collaborative enterprise, a student is judged individually for her accomplishment, and that judgment often transpires publicly, where her project is deemed admirable, promising, or wanting. The physical configuration of the trays, where every person's work is always on view to everyone else, enables and reinforces this fiercely every-student-out-for-him-or-her-self atmosphere.

The configuration of the trays exacerbates the problem in two additional ways. The trays' vertically stacked arrangement reinforces the atmosphere of competition by physically asserting a manifest hierarchy: as students move from first to second year and so on, they literally take their place above the more junior students, moving to the higher-up trays. And on all but the lowest tray, the drawing tables are arranged in long, narrow lines, affording students many fewer opportunities to congregate informally in small groups than if their workspaces were more conventionally shaped. Long, linear spaces suggest avenues of movement, and they discourage precisely the kind of happenstance social gathering that nourishes community.

Fortunately, the almost-square space of the café, located in sight of the trays and offering ever-changing views to an enclosed yard behind, provides an antidote. Food offerings and cashier stations line its perimeter, with movable tables and metal chairs filling the room. This is an immediately recognizable, patterned arrangement indicating that this place affords leisurely socializing, eating, hanging out. Through the interactive dynamics of our previous experiences with this and countless other kinds of action settings, we know what's done here. We "intuitively" *know* that long rows of desks facilitate quiet activities and rows of long tables with clanking chairs don't. Just a glimpse or even the mere thought of one type of action setting is enough to bring such schemas to mind.

As our experience of the GSD illustrates, the built environment is one of the principal ways that we as individuals apprehend, experience, participate in, imagine, inculcate, and retain the social world's norms, conventions, meanings, and possibilities. How designers manage affordances, create action settings, and communicate their character influences the kinds and qualities of attachments we develop to both people and places. In the past decades, a scattered but growing number of architects, theorists, and psychologists have begun to delve into the cognitive dimensions of built environmental experience, but few have brought the same kind of analytic rigor to understanding how our individual, embodied experience is inflected by our situation in the social world. Just by living in our bodies in the places of our shared world, we develop a vast storehouse of schemas connecting built spaces to social grouping and actions, and it is within this context that we are constantly forming and executing (or getting distracted from) our goals, so the places we encounter should be designed accordingly. Last, since we cannot but imbue places and things with meanings—nonconsciously and consciously—the compositions of our constructed worlds must spark appropriate emotional and cognitive associations.

What's Next

Design is a social instrument. Built environments shape social relations. This is true for every place that we live and everywhere that we go. So people's experience of the built environment is at once private and individual—situated in our bodies and the natural world—and public, situated in our social worlds. Here at last we have a conceptual framework that offers a complete descriptive account of the recursive, constantly interactive character of the relationship among the individual human mind, body, and social environments in which we live.

What remains to be explored is how the spaces and places that individual minds and bodies, within social institutions, design and construct could be made healthier, more vibrant, more inspiring communities and societies. What are the design principles and tenets and what are the social ideals, that will constitute the standards by which we judge our built environments, today and tomorrow? Now finally we can begin to construct an answer.

CHAPTER 6

Designing for Humans

A building and the space it possesses should help us be alive, it should allow for the heeding of things. . . . Is there a difference between making events possible and creating them?

—ANNE MICHAELS, *The Winter Vault*

What we now know about how natural and social environments affect people suggests that for the design of our landscapes, urban areas, and buildings, some orienting guidelines will help prevent easily avoidable mistakes, promote design that better satisfies people's needs, and promote human well-being. The immense complexity and plasticity of the human brain and the extraordinary richness and cultural and geographic variability of human experience will guarantee that experientially informed aesthetic principles could never result in overly formulaic design. Instead, such principles will free designers up and permit them to explore the plethora of compositional possibilities they allow, while maintaining the human-centered approach that constitutes experiential design.

Human embodiment itself elicits initial tenets, most of which are directed to our nonconscious selves. Places should be scaled to the human body—but in both its egocentric and allocentric manifestations. We apprehend our environments in ways that are imaginative and constructive, and so they should be designed to work with the minds that humans have, eliciting a range of apt associations. We also experience our built environments with all of our sensory faculties working in concert, and with all of them collaborating with our motor systems for action. So in this dimension too, design must proceed accordingly.

Legibility in overall forms, especially in larger constructions—like the lilting forms of Frank Gehry's Guggenheim Museum Bilbao or the hard-edged, prismatic shapes of Florence's Piazza del Duomo, with its Baptistery, bell tower, cathedral facade, and dome—works well with our rapid-scanning, gist-extracting eyes and brains. But our yearning for legibility must be balanced with our craving for cognitive stimulation, and legible constructs need not be simple ones. The Salk and the National Pensions Institutes taught us that whatever a project's overall configuration, its surfaces (including temperature, pliability, color, density, and so on), its materials and their textures, its auditory qualities and more will greatly influence our sensory, cognitive, and especially our emotional responses to them, and in doing so shape our experience of them. Along with carefully chosen materials, a well-designed environment contains well-conceived, well-executed construction details. These contribute a sense of scale, adding visual (and sometimes conceptual) depth to a project by tickling our sensorimotor imaginations and promoting our cognitive engagement with the places we inhabit, and with the objects in it. Amiens Cathedral, Museum at the Stream, and other projects show that nonvisual cues—auditory, tactile, proprioceptive, olfactory, and so on—contribute in essential ways to our overall experience of place. Nature's bounty—natural light, greenery, awareness of site and climate—should always be an essential consideration in design, and designers can learn myriad lessons from nature, simulating or abstracting its forms, managing its climatological, topographical, and material particularity in design. These are the basics; in addition, some other dimensions of experiential design need parsing in greater detail.

Ordering Patterns: Embodied Math, Embodied Physics

Patterns, which people rely upon to differentiate a place or structure from its surroundings, advance legibility and can create coherence. Humans are ever on the lookout for iterative patterns because the very machinery of our sensory cognitive systems—our propensity for rapid gist identification, the goal-oriented nature of perception, and our susceptibility to primes—requires us to first, quickly parse foreground from background, and second, assign meaning to the things we encounter. Recognizing and identifying patterns produces in us the sensation of pleasure. Whether it's when we listen to a piece of music or look at a painting or walk through a building or landscape that slowly reveals the nature of its order, recognizing patterned organization rewards us with a little jolt of the opioids in the area of our brain associated with our "liking" system. Presumably, the functional origin of this reward system lies in our evolutionary need to rapidly situate ourselves and the members of our group within an environment and a social group.

That humans powerfully gravitate toward legible environments is illustrated by how consistently we recoil from ones so complex that they

Without patterns, chaos: Rehak House (Coop Himmelb(l)au), unbuilt

deflect our attempts to extract their gist. In the 1990s, for a brief period, a group of European and American design practitioners fell under the influence of the French postmodern notion (originating in the writings of Michel Foucault and Jacques Derrida) that any ordering system rests on a foundation of randomness, illogic, and the institutional exertion of power. A handful of architects in Europe and the United States espoused "deconstruction"—which had originated as a technique of literary and philosophical criticism—as both an analytical heuristic and an artistic technique. Among the ensuing experiments with ostensibly disordered buildings was the Rehak House by the Austrian Coop Himmelb(l)au (the firm's name means both "studio blue sky," when the *l* is included, and when the *l* is omitted, "studio sky build"), whose principal architect, Wolf Prix, once told an audience that he had derived the design for one project by sketching what he recalled of a dream with his eyes closed. But such designs frustrate our instinctive perceptual strategies, repelling any effort we make to orient ourselves in relation to them. We wonder: What might be this object's purpose? Is it relevant to my or anyone's life? Where would I enter? Would my experience of it be aggravating or disorienting and frustrating, or rewarding? Not surprisingly, neither the Rehak House nor most of Coop Himmelb(l)au's comparable early projects were built: theatrically disorganized compositions violate people's experiential needs. Patterns are a necessary component of any built project.

Throughout history, designers have relied on mathematical systems and the principles of physics to establish both visual and structural frameworks for the environments they build. Often, especially in premodern societies, the composition's pattern was driven or even determined by the structural properties of its materials. When the ancient Egyptians constructed the majestic Mortuary Temple of Hatshepsut, for example, they arranged its forest of columns in closely spaced rows. Structure determined form: stone is strong in compression but weak when stretched (which puts it "in tension"), so the Egyptian masons knew that the lintels they used to create internal spaces could not stretch across large spans.

The physics of materials alone is rarely enough, though, to determine the whole of the composition's patterned structure. Other mathematically grounded schemas are also deployed; of these, the most common are simple symmetries and Euclidean solids (the latter resonating with our large, internalized alphabet of geons). Simple Euclidean solids can

Geons in buildings

be found everywhere, from the north and south facades of the Salk Institute to the volumetric organization of a local bank. Symmetries too can be simple or complex. While the bilateral symmetry that guides the composition of the Salk's central plaza is straightforward, the best-known example of a more complex kind of geometry is fractals: apparently irregular, iterative compositions that display individual branching structures that repeat at many different scales. We can see fractals in nature—as in the structure of a shoreline, a fern frond, a calabrese romanesco cauliflower—and in culture—Gothic cathedrals display fractal organizations, as do Hindu temples in an even more comprehensive form. In the eleventh-century Kandariya Mahadeva Temple in Khaju-

Fractals in buildings: Kandariya Mahadeva Temple, Khajuraho, India

raho, India, fractals govern the articulation of the building's surface elevations, the relationships of each of its shapes to one another (and therefore, its floor plan), and the visual and spatial hierarchies that establish its proportions. Because not only fractals but also symmetries and Euclidean geometries appear in nature, some argue that our evolution over tens of thousands of years has made us constitutionally attuned to seek out and take pleasure in such mathematically guided compositions and proportional relationships.

How does embodied mathematics and physics inform the ordering patterns upon which we rely, and what kinds of experiences do they offer? An iconic example can be found in the best-known buildings on

Gateway (Propylaea) to the Acropolis with the Parthenon in the distance, Athens, Greece

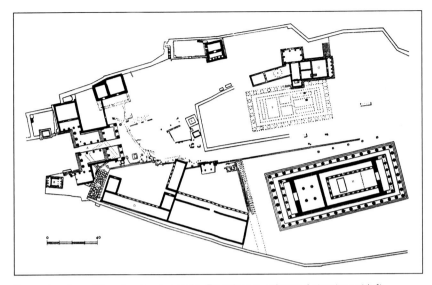

Reconstruction of the site plan, Acropolis: Propylaea is columned structure at left, Parthenon is bottom right

the Acropolis in Athens, which were constructed in a single campaign under Pericles, during the fifth century BCE. Standing at the gateway to the temple precinct, the Propylaea, we immediately notice the patterned order of this hilltop temple complex, even in its ruined state. The Propylaea's fluted, baseless Doric columns establish the visual theme that unified this gateway structure with the other major buildings on this site. Constituting the facade of the Propylaea are a paired set of three evenly spaced columns disposed around a wider, central void, which opens the way to our path forward. At the same time, as we move around inside the Propylaea, the range of columns screens and frames one of the most famous prospects in all of western architectural history: the majestic Parthenon rising diagonally to our right, above us in a three-quarters view. To our left, farther in the distance, stands the sublime Erectheion, the complex, multi-prismed, asymmetrical building that presents four graceful female statue-columns, caryatids, rising to support the porch's pediment with apparent ease.

Caryatids function as columns: Erectheion, Acropolis

The regularly spaced columns on the hilltop Acropolis satisfy our human pattern-seeking wiring. The simple rhythm of a Doric colonnade—solid, void, solid, void, solid, void, solid—enables us to apprehend the coherence of the complex and begin to make sense of it as a whole. At the same time, the variation of the column surfaces—with sculpted fluting that exaggerates their verticality and hides the seams between each masonry block and the next, and with their dynamic tapering from bottom to top to illustrate the force of gravity pressing down through them, with the transformation of column to caryatid on the Erectheion—makes that columnar pattern complex enough that it refuses to recede into our visual or conceptual background. The columns' design, size, and intercolumniation (the dimension of the spacing between columns) varies, from Propylaea to Parthenon to Erectheion, simultaneously individuating the buildings and establishing a theme common to them all.

Regarding the Parthenon from the Propylaea, we see a rectangular prism, diagonally disposed. Its placement on the site in perspective view practically insists that we approach it. Why? If this were not a ruin but the completed building, we would in short order identify the door leading to the Parthenon's indoor room, the Cella, at the midpoint of the prism's short end. An ancient Athenian would have known that the priest was the only person permitted to enter the Cella; even so, this aperture's very existence (as well as the narrative suggested by the Parthenon's pediment sculptures, long gone) would have produced a tension between our initial, diagonal view of the temple and the frontal view that this entrance suggests. Resolving this tension requires *action*: we must transport our bodies until we reach an approximately frontal vantage point.

The majesty of the Parthenon lies partly in the sublime order it exhib-

Approaching the Parthenon, Acropolis

its, partly in the precision of its symmetry. Humans are strongly drawn to bilateral symmetry, at least in objects like a single building. This might have something to do with our goal-oriented approach to environments and our reliance on rapid gist identification: symmetry's iterative nature, with one side of the object mirroring the other, is satisfyingly predictable as well as easy to navigate. It could also be that we have become habituated to symmetry owing to its ubiquity in both nature and the built environment. Or, if we use what we know about the development of human cognition, people may be drawn to symmetry in objects because it duplicates our experience of how people's bodies are disposed.

The most important category of things in the world for people are other people, and the overall shape of the human body and its face is symmetrical along a vertical axis. The appeal of bilateral symmetry does appear to be innate: even very young infants gaze at such objects longer than they do at asymmetrical ones, and this is true across cultures. "Good symmetry," neuroscientist Eric Kandel writes, "indicates good genes"—and, he might have added, robust health. Even without our conscious awareness, our evolutionary heritage has taught us that almost every healthy animate being exhibits symmetry either globally, in its overall composition (the form of a butterfly) or locally (the pattern on its wings) or both. Symmetry in a perceptual object, then, heralds (in the words of V. S. Ramachandran) a "biological object: prey, predator, member of the same species, or mate." Although the objects in the built environment, including its buildings, are inanimate, symmetry may also appeal to us because it intimates a human presence.

People, in their bodies, nonconsciously are likely aware at all times of the forces of gravity. We also perceive symmetrical compositions as "balanced," just as, when we stand on two legs, we stand "firmly planted" on the ground. In this case, visual symmetry complements the perceptual schemas we develop from our embodied knowledge of gravity and the properties of physics.

People see bilateral symmetry in the built environment as both good and bad. When urban designers and architects arrange bilaterally symmetrical large-scale buildings or complexes—places so large that we tend to perceive them as *scenes* rather than as *objects*—people's reactions are not consistently positive. Consider the palpable stasis we feel at Mansudae Assembly Hall in Pyongyang, in North Korea, where repetition and bilateral symmetry confer the impression of soporific, dehumanizing pattern and oppressive control, and compare that to the dynamism of our experience of the site plan of the Acropolis, where we proceed from the Propylaea to the Parthenon or the Erectheion.

At the Acropolis, most of the individual buildings convey an impression of symmetry, even though neither the Propylaea nor the Erectheion are precisely symmetrical in and of themselves. And all the buildings at the Acropolis are asymmetrically arranged on the site. The logic of their placement eschews simple math and takes cues instead from the embodied physics of our place on the ground and our move-

Mansudae Assembly Hall, Pyongyang, North Korea

ment through the topography of the hilly Acropolis site (as well as from the preexisting sacred altars on the ground, which partly explains the intricate complexity of the extraordinary Erectheion). As we explore the buildings and spaces of the Acropolis, the palpably productive tension we feel as we move around and experience the site results from the conflict of its two ordering systems: mathematically regular tropes in its buildings, in contrast to the asymmetrical, mainly topographically specific and physics-based principles governing the arrangement of the buildings on the site.

Our introduction to the Parthenon offers a view of both its front and lateral facades, demanding that we imagine it in three dimensions, as an object in space. This enhances our sense of its weighty mass. These buildings, after all, were constructed from the huge blocks of masonry that remain today, scattered around the site. Because we live in bodies and in all likelihood have experienced, at some point, the weight of stone, we nonconsciously know that these masonry blocks must have been very, very heavy. And because the depredations of time and humans have not dismantled these ancient buildings, their deteriorating state prompts us to think about the people who constructed these monuments, and how they must have struggled to complete them. Our past experience in our bodies, struggling with gravity's forces, suggests that this endeavor could only be a human cry against mortality: the people responsible for this complex designed it in such a way that the prospect of demolishing it was more onerous than simply bestowing it to the stewardship of time. The Parthenon's immensity and symmetry, as well as the Acropolis's rhythms and variations, bespeak the victory of human artifice over the vagaries of nature in this craggy, uncompromising site.

The complex interplay of embodied math and embodied physics creates a sense of rightness about the Parthenon's disposition. It harmonizes with its surroundings as much as it dominates them. This may be owing

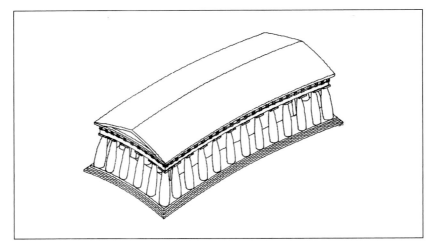

Illustration exaggerating the optical refinements in the Parthenon

in part to an additional property of the building that is not immediately visible but may be sensible on a nonconscious level: the optical refinements. The Parthenon looks to be full of straight edges. It's not. Not only do its columns taper from bottom to top, expressing the downward pull of gravity and the accretive weight of the stone; they also bulge slightly at their midsection, in a refinement known as *entasis*. They also lean slightly inward as they rise toward the lintel. And the platform-base of the building more resembles a pillow than the flat plinth it appears to be.

Powerful design reasons must underlie the inclusion of these optical refinements, because they would have made an already ambitious design devilishly difficult for masons to construct. Perhaps the building's architects, Ictinus and Callicrates, intuited that people respond more positively to curved surfaces than straight ones. Perhaps they understood that repeating straight-edged columns and planes on such a large scale might be perceived as inert and oppressively static. It's likely that Ictinus and Callicrates used the optical refinements to further imbue the building with a palpable sense of movement: subtly, they make the

Parthenon look as if it were heaving, responding to its own immense physical bulk and weight.

The Parthenon, built in honor of Pericles, one of the ancient world's celebrated leaders, and dedicated to one of ancient Greece's most important goddesses, Athena, is a temple designed to elicit specific emotional responses—awe and inspiration—and along with them, a specific set of cognitions relating to the rectitude of its builder's political and social ideals and the inevitability that the institutions he built would endure. It and the buildings surrounding it on the Acropolis were meant to make those ideals seem to be both born of nature and superior to its vicissitudes. To the extent that these buildings succeed, they do so *by design*. Through patterns established by embodied mathematics and physics, the entry sequence leading to the Parthenon, the buildings surrounding it, the temple's placement on the site, its size, its materials, and its design all manage our experience by working with our sensory and motor systems to captivate and sustain our attention and imprint themselves on our imagination.

Complementing Patterns with Complexity

Patterns in the absence of complexity repel us. A look at typical developer-built tracts of residences is enough to know that sameness and repetition dull the senses. That is why generations of writers have deemed the German architect Ludwig Hilberseimer's proposal for the modern city (1944) horrifying. More recently, the Danish urbanist Jan Gehl found that pedestrians walking in cityscapes are happiest when they find something new and interesting to look at approximately every five seconds. This is also why the radical simplification of form advocated by Hilberseimer's colleagues—early modernists—produced a built vernacular that was (unjustly) vilified by people around the world.

So within designs, pattern must be coupled with complexity. Take, for example, one pattern that is commonly found in the built environment, simple repetition, as in grids or classical colonnades. On a small scale, as in the Acropolis's tiny Temple of Athena Nike in Athens, the simple light-shadow-light-shadow repetition makes an ionic colonnade looks lovely. But in the US Treasury in Washington, DC, the repetition of simple patterns repeated seemingly ad infinitum, at a very large scale, becomes boring or even downright enervating. This is owing to our human craving for cognitive stimulation, as important to us practically as oxygen is to us physically.

Below: The Federal Treasury, Washington, DC

Opposite: Temple of Athena Nike, Acropolis

In addition to patterns, then, what people need from their built environment is respite from patterns. Another way to conceptualize this is to say that what people need from the built environment is *patterned complexity*, which is a structuring principle derived from organic form. Built environments exhibiting patterned complexity offer stimulating landscapes that we can return to repeatedly, knowing that they will enhance our moments and days.

Patterned complexity, as a general principle, allows immense latitude for creativity, both because it embraces a wide range of design approaches, and because it can guide the composition of a wide range of elements, including volumes; spatial sequences and views; materials; sonic, thermal, tactile, and other properties; and the ways in which the constructed object relates to its site and the natural world. Compositions exhibiting patterned complexity offer manifold experiences. Some function as arenas of serenity, allowing our minds to freely wander, replenishing our depleted attentional and other internal resources. Others force us to actively problem-solve, honing our cognitive capacities. Both leisurely exploring and active problem-solving constitute modes of cognitive stimulation that people need and benefit from every day.

The most primitive form of patterned complexity involves chunking spatial volumes in accordance with either the structural possibilities of a construction's materials or the various action settings within an institution, or both. These can coincide with one another, in which case visual or volumetric chunking quickly reveals both a project's underlying structural and its functional organization, and to apprehend its meaning, we immediately refer to our "categories are containers" schema, which prompts us to identify the action settings housed in its volumes—kitchen here, bedrooms there, and so on. Chunking appears frequently in ancient architecture, as in the Temple of Hatshepsut and the Acropolis, when the available technologies and materials limited spatial dimensions. It appears also occasionally in contemporary projects, as in Gehry's Schnabel

House in Los Angeles, where the living room, the kitchen, and sleeping areas are housed in individuated containers.

Because of our cognitive needs for both order and complexity, the most successful patterns are ones where a clear theme is introduced, then progressively varied, as in a musical composition. At Harvard, in Gund Hall's large porticoed facade, the stripped classicizing colonnade is built up from the two ranges of columns that are spaced differently. This itself creates a complex rhythm, for which the window-walls supply a contrapuntal variation. Theme-and-variation compositions similarly regulate nonclassical designs, as in a small library by David Adjaye. Structurally and functionally, Adjaye's Francis A. Gregory Neighborhood Library in Washington, DC—a public library built on a tight budget—is little more than a large rectangular shed with exterior walls, patterned out of a

Chunking by function or structure: Schnabel House (Frank Gehry), Brentwood, California

Francis A. Gregory Neighborhood Library (David Adjaye), Washington, DC

Patterned complexity in the interior of Francis A. Gregory Library

simple diamond-shaped grid, or *diagrid*. To this diagrid datum, Adjaye introduces complexity in three ways. He alters the shape of the windows on different facades by "pulling" the grid this way, then that. In some places, the windows become squares turned on end; in others, they morph into elongated, attenuated diamonds. He also stitches the building's diagrid facade into a patchwork quilt of reflected surroundings with mirrored surfaces that change along with the weather. And inside the building, those shiny flat diamonds turn into fleshy three-dimensional cubbyholes that overlap this way and that, becoming shelves, and even, on occasion, inglenooks in which to perch. A simple structure, the Francis Gregory's surfaces, materials, and skin that pack the building's experiential punch poetically illustrates how even in projects with relatively modest budgets, surface-based cues drawn from texture, materials, colors, and construction details can be creatively implemented to shape a profound emotional and cognitive experience.

The term "patterned complexity" may bring to mind highly ornate, deeply sculpted surfaces, yet simply articulated, patterned frameworks can make captivating places at any scale just by orchestrating surfaces to display appealing variations. In the gridded office slab of Sauerbruch Hutton's GSW Headquarters in Berlin, plate-glass windows sit in a straightforward metal framework, while tonal variation in oranges, beiges, pinks, and rose-red hues—the window shades—differentiate this governmental office building from its surroundings, and allow the office workers inside to identify by color with the particular part of the exterior as that place they occupy inside, helping them to develop a strong sense of place.

In the tiny St. Nicholas Orthodox Christian Church in Springdale, Arkansas, Marlon Blackwell used surface-based cues to transform a generic three-bay, aluminum-sided garage into a jewel of spiritual contemplation and worship. Working within severe budget constraints, Blackwell chose to retain the overall form of the garage and its box rib

metal panel siding. Then he offset the exterior's repetitive vertical stria-tions by affixing a projecting portico over the front entrance and asym-metrically placing three recessed apertures at the right and left edges and above the doorway of the building. He also dug a band around the building's perimeter, filling the trough with smooth, touchable, hand-sized black stones, thereby differentiating this shed from its surround-ings in a way that, from a distance, makes it appear to hover slightly above the ground.

A deft use of color enlivens the entry sequence. In the newly in-stalled tower, the cross-shaped window's tinted glass washes red light over its deeply saturated red and white walls; opposite this entrance,

Below: St. Nicholas Orthodox Christian Church (Marlon Blackwell), Springdale, Arkansas

Opposite: GSW Headquarters (Sauerbruch Hutton), Berlin, Germany

a daffodil-colored concrete stairwell sings a brilliant yellow light. The pulsating hues of the circulation spaces render the white sanctuary, with daylight washing down its eastern wall, that much more serene. A shallow dome—before Blackwell's recycling, it began life as a satellite dish—anchors the center of the room. Near the altar, a metal scaffolding displays icons of saints with their attributes; these surround a painted depiction of the crowned Jesus, hands raised as if bestowing grace on all who enter.

The Francis Gregory Library and St. Nicholas Church reveal how patterned complexity can transform even small, very modest buildings into visually and emotionally intense places. As a design tool, patterned complexity is suppler still, because designers can subvert a project's dominant pattern through nonvisual means. The St. Nicholas Church exemplifies this as well. The building's exterior introduces a banal sight so familiar in the American landscape that a split second's glimpse primes us with a cascade of associations—garage. Prefab. Chintzy. Cold. Yet here is a *beautiful* place, with clean lines, carefully executed construction details, deliberately planned and managed spatial sequences, and color-saturated surfaces. The building's emotional impact is owing not only to *literal* variations in its surface patterns, but also to the ways that its execution and details violate our initial *expectations* about what sort of action settings this building must house. Blackwell introduces complexity to this patterned object through defamiliarization: suddenly, something familiar reveals itself to be strange, and thereby captures our imagination.

Designing Change into the Built Environment

Habituation is the biggest obstacle to built environments that enrich our lives and places. Unmoving nonthreatening familiar objects and surroundings do not capture our attention. Even elements of good

Foyer, St. Nicholas Church

Sanctuary, St. Nicholas Church

design can dull our senses over time. But designers can forestall habituation in the built environment and mitigate its soporific effects by taking advantage of the changeability of nature and people's use of actions settings.

NATURE'S TOOLS

Like Andy Goldsworthy's *Sticks Laid One Way and Another to Turn Dark to Light and Light to Dark as the Sun Rose and Set*, places can be designed in such a way that they are so responsive to environmental variations—light, weather, temperature, and sound—that they *appear* to be changing even when they are not. In Denver, the wonderful Clyfford Still Museum (CSM) by Brad Cloepfil Allied Works Architecture exemplifies this approach. A two-story, exposed-concrete prism, the CSM instantiates a radically simplified visual language. Even so, its highly textured surfaces include seemingly raked, "corduroy"-concrete walls that cannot help but activate our sensorimotor engagement and rebuff our tendencies to glance over them because they change all day as the sunlight cast inside shifts position with the weather and over the course of the day. Outside, planes of corduroy-concrete, along with

Below: The interaction of materials in light: Andy Goldsworthy, *Sticks Laid This Way, Then That* (detail), photographed at three different times in one day

Opposite: Materials and texture in light: Exterior wall detail, Clyfford Still Museum (Brad Cloepfil), Denver, Colorado

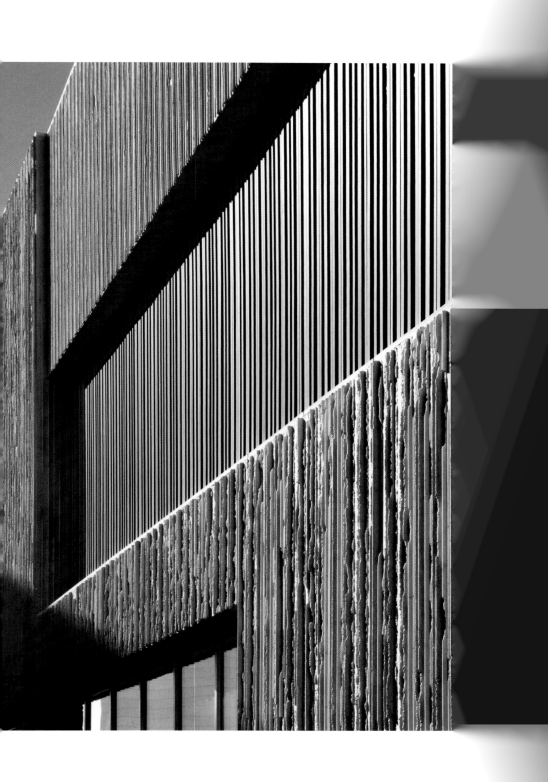

flatter passages imprinted with the texture of wood grain, recess behind and protrude in front of one another. Inside, these surfaces abut ranges of darkly stained wooden slats. The result? The CSM becomes a muscular shadow box, surfaces shifting in repetition and variation, exploding into bright, filtered light when we reach the main gallery spaces on the second floor. Here these highly modeled, straight-lined surfaces set off the diagonally raked, deeply set grid in the ceiling, with its stretched, oblong-shaped skylights. Such textural variations in the context of the CSM's visual simplicity draw us into an embodied relationship with the building. Because its striated surfaces shift in hue and model light and shadows all day long, Allied Works manages to exaggerate the building's very stasis while highlighting the exterior world's changeability. Such a place sets the stage for our heightened awareness of moment-to-moment incidents. Refusing to quietly retreat into the background of our awareness, operating on all of our senses, they insistently elbow their way back into the realm of our conscious attention.

Change can be designed into the built environment not only by us-

Materials, light, geometry, texture: Interior, Clyfford Still Museum

ing natural light to highlight texture and the passage of time, but also by featuring greenery, climate, and the topography of a particular (and especially a distinctive) site. Unlike a building's stasis, these natural features change over time, and sometimes even from one moment to the next: grasses flourish and wither, weather blows with the wind from dawn to dawn. Even the very earth—a site's topography—shifts and erodes with the passing of eons and days. As a result, projects which feature or exaggerate nature's presence will change—with the weather, over time.

The Sea Ranch, a 4,000-acre development of vacation homes on California's Pacific Coast, located a hundred miles north of San Francisco, accentuates the experiential benefits of designing with nature's botanical and mineral changeability. Designed in 1965 by a team of architects, landscape architects, a real estate developer, and a geologist, this complex remains regulated by covenant. At its fiftieth anniversary, the Sea Ranch still constitutes one of America's best arguments for intelligent design review. The Sea Ranch's simple, pitched-roof houses—most of modest size—feature unpainted, weathered Douglas fir or redwood cladding. Arrayed along the Pacific coastline, the structures are placed behind a walking trail for public use that is perched on the crest of the ocean's dramatic cliffs. This site plan ensures that the Sea Ranch's main events will always be the undulating, eroding, windswept landscape, with its ever-changing native purple, yellow, and wheat-colored grasses; its succulents and ground-hugging shrubs, spread wild along ten meadowed miles. Despite or because of its retiring architecture, the Sea Ranch captivates our attention and soothes our senses, corralling us into its aegis again and again. Nature, its changeability, and how people must defer to its laws when inhabiting the land create the Sea Ranch's compelling beauty. As the sun moves across the sky, cumulus clouds cast dark moments along the plain; shadows fill, then take leave of porches, gables, and the occasional turrets. The sharp morning light

Typical house, the Sea Ranch

etches the buildings' edges into the spreading landscape; later in the day, softly cascading pools of dusk cast an even glow. Nature plays its windy game against the patterned scatter of these modest, weathered homes, carefully engineered to appear untouched, which themselves transform the human act of building from a stubborn statement of defiance against impermanence into a fragile, tentative hypothesis of affiliation with the natural world's endless variety and monumental beauty.

More and more contemporary designers, with and without the Sea Ranch as an example, are sculpting from nature's changeable features magical theaters of place. Alvaro Siza's Leça Swimming Pools in suburban Porto, Portugal, present a powerful public landscape-cum-architecture project dominated by its craggy, rock-strewn site. Vo Trong Nghia's prototype for affordable housing in Vietnam, the S House series, specifies structural frames in reinforced concrete or steel, but the selection of its exterior cladding is left to the construction exigencies

Changes in climate activate the site: view from the ocean, the Sea Ranch, California

Leça Swimming Pools (Alvaro Siza), Porto, Portugal

and available materials of the Mekong River Delta: bamboo and nipa palm leaves are variously, locally sourced, saving shipping and construction costs while embedding projects into their immediate locales.

ACTION SETTINGS AND PEOPLE

A second effective way for designers to mitigate the built environment's stasis takes advantage of the vitality of its action settings by making the human body's presence and movement in space the animating features in a design. Wright's Guggenheim Museum in New York and Gehry's Guggenheim Museum Bilbao are famous examples. In New York, a spiraling interior opens views across its central atrium onto other artworks and other museumgoers, who are themselves engaged in art-watching and people-watching. Before Wright, Charles Garnier, the French nineteenth-century architect, multiplied the number and expanded the range of available action settings in his foyer of the Palais

Garnier Opera House in Paris. Reconceptualizing the experience of a night on the town, Garnier transformed the opera house's entrance, corridors, and staircases into something of a second stage: a public stage of self-presentation as the antecedent to the private stage of a theatrical performance. Entering the opera, we face the curving sweep of lavishly sculpted shallow steps, rich in marbled reds and greens. The staircase splits into two curving ranges as it rises to meet the surrounding foyer at opposite ends of the piano nobile. A straightforward space of transition is transformed into a lively antechamber for the public, where operagoers preen and mingle, playing their own self-appointed roles in the public spectacle that precedes the performance.

Individual bodies animate interiors on a smaller scale in Rem Koolhaas's brilliant early buildings, both the Villa dall'Ava in a suburb of Paris and the Kunsthal in Rotterdam. The Villa dall'Ava's residents become subjects in a carefully framed, almost *Rear Window*–like series of

Wright envisioned the Guggenheim Museum as a place where people would gather to look at art and each other—while the kids entertained themselves

The Palais Garnier foyer, with multiple staircases and landings, was designed as a place to see others and be seen (Charles Garnier), Paris, France

vignettes: walking around the house and the site, we might catch sight of the daughter's private moments through the large plate-glass window in her room. At Kunsthal, as we take in the art while standing in the lower galleries, dramatically foreshortened bodies of unknown gallery-goers suddenly appear above us through the translucent ceiling-floor plane separating the stacked gallery spaces. We become unwitting actors on someone else's stage just by the very act of inhabiting such places. Koolhaas's irony-drenched approach to design, titillating and surprising, has influenced legions of subsequent projects, including New York City's High Line by James Corner Field Operations with Diller Scofidio + Renfro, where an abandoned, elevated railroad trestle was transformed into a one-and-a-half-mile-long public park that slashes its way in uneven lines from West Midtown, down through Chelsea, and into the Meatpacking District. Walking along the High Line leads us around and sometimes

Villa dall'Ava (Rem Koolhaas/Office of Metropolitan Architecture), Paris, France

through the bellies of the older warehouses and newer residential buildings of West Manhattan. Catching glimpses of the Hudson River, we stop to rest, lounge on one of its many chaises, or join others in one of its amphitheater-like seating areas. It's a compelling place that transforms us, in our bodies, into sculptures posed on pedestals, and frames small social groups, presenting them to city dwellers in picturesque scenarios. Animating environments by stage-managing bodies—people—as Garnier, Koolhaas, and James Corner Field Operations do constitute a powerful counterargument to the poorly designed escalators and other lost opportunities that bedevil so many of the world's buildings, including even its most prestigious museums. Captivating, active, body-mindful public spaces combat over- and understimulation by offering intriguing, vibrant places that feature an ever-changing cast of people and their movements as focal points in the composition.

The benches at the High Line turn people into part of the composition (James Corner Field Operations with Diller Scofidio + Renfro), New York, New York

Character: The Puzzle of Well-Chosen Metaphors

Places that intrigue and engage us, that draw us in and draw us back, come imbued with *character*. While the notion of character might seem amorphous, in this context it means something quite precise. Character emerges when an isomorphic relationship exists between a place's literally designed forms and its action settings. An institution acquires character through design, through the manipulation of our biologically wired, direct responses, such as our human disposition to approach curving surfaces and retreat from sharp-edged ones, and to intuit how objects behave according to the principles of gravity. A place is imbued with character also through the embodied schemas and metaphors it elicits, which express or support an institution's objective functions or its broader social ethos. Metaphors impart character because people comprehend them by conjuring up mental simulations of their referents, which carry with them associated emotions and actions. To take a simple example, for a private home study, the architect Nader Tehrani sculpted plywood bookends so that they resemble turned-out books. The bookends themselves elicit thoughts of books on shelves, and perhaps also of the wood that constitutes both the shelves and the pages of a book. Such associations add semantic density to what is, functionally, a simple range of shelving.

Metaphors add character to places—but only when they are chosen well. If a metaphor is too abstract, the design risks incomprehensibility. If a metaphor is overly literal, the composition risks devolving into a one-liner which people will apprehend quickly and then rapidly habituate to. Jørn Utzon's Sydney Opera House exemplifies the delicate dance a designer undertakes to construct an effective metaphor. The opera house's vaunted, vaulted forms evoke associations of half-buried shells projecting from the sandy beach; or a giant, ossified Triceratops or other prehistoric creature; or billowing "sails" that rise from the anchored platform of a skiff, catching the breeze at a prominent point in the city's harbor. The Sydney Opera House's multivalence constitutes part of its experiential power. It elicits varied and different metaphorical associations—beach, water, wind, movement, soaring upward—which distinguish this harbor city and showcase the experience of the opera house's action settings: listening to music, transformations of the wind.

In contemporary architecture especially, designers rely heavily on embodied metaphors, which have become a principal means of creating enriching, emotionally engaging public places. Herzog & de Meuron's Olympic Stadium in Beijing alludes to a bird's nest (which is also its nickname), with all the associations of delicacy, fragility,

Metropol Parasol: trees (or giant mushrooms) shading a large plaza in central Seville (Jurgen Mayer H.), Spain

and ephemerality of aviary creations. But exploded in scale and re-made in steel, their "Bird's Nest" becomes a monumental arena for public spectacle. Frank Gehry's 8 Spruce Street, a seventy-six-story residential, metal-clad skyscraper, elicits thoughts of a soft, shimmering curtain hanging from the sky amid the towers of New York City's financial district. And Jurgen Mayer H.'s bonded timber Metropol Parasol is a biomimetic enclave of coffered trees (or giant mushrooms, as Sevillians prefer) set in the historic center of Seville, shading the Encarnación Plaza.

Carefully devised, skillfully deployed metaphors mitigate the built environment's stasis and our tendency to habituate to it, through the many overlapping associations they elicit. They reward repeat visits,

Sails, shells, and soaring: metaphors at the Sydney Opera House (Jørn Utzon), Sydney, Australia

Bird's Nest Stadium (Herzog & de Meuron), Beijing, China

sparking new thoughts and fresh emotional experiences over time because the difference between the referents to which these metaphors
allude—smooth sailing on a great yacht, cuddling in a cozy nest—
and what these buildings actually are—an opera house, an enormous
stadium—never disappears. In a constructed metaphor there will always be a distinction between its obvious referent—wind-catching
sails, bird's nests, fluttering curtains, the spreading canopies of trees—
and its physical actuality—tile-clad concrete vaults, huge crisscrossing
concrete piers, stainless steel, wood panels arranged in curving coffers.

That distinction, ever present, will not be forgotten. The Sydney Opera House's concrete vaults will never become fabric or shell-like, and the spreading wooden canopy of the Metropol Parasol will never grow leaves or be plucked for a tasty stew.

Such meaningful moments transform buildings—which tend to retire into our experiential backgrounds as *scenes*—into *events*. They corral us into time capsules of heightened cognition because our nonconscious expectations are contravened, and momentarily we become estranged from the ordinary rhythms of daily life. The background expectations which the Sydney Opera House overturns, by dint of its very presence, are many: that most buildings and constructed places are arranged orthogonally; that what resembles a sailboat *is* a sailboat; that buildings neither look nor function like sailboats—in fact buildings are quite unlike sailboats, as they need to be firmly anchored in the ground. And yet on Sydney's waterfront, here we have it: a nonorthogonally arranged opera house that loosely resembles a giant sailboat.

Coming upon such puzzles results in a pleasurable sense of cognitive engagement. Resolving the puzzle demands creativity and focus, and because the solution is never obvious, we keep working at it over and over again. What's with these huge folded vaults in Sydney's harbor? Why does this immense stadium in Beijing elicit thoughts of a fragile sparrow's nest? Why does a church in Arkansas resemble and initially elicit associations of a work shed or a garage? The "click" of realization we feel with an explanatory insight corresponds neurologically to a release of pleasure-inducing endorphins in the parahippocampal cortex, and that pleasure, combined with our cognitive engagement with the project, becomes encoded as part of our experience of that place. Such places are more likely to "stick" in our memory, becoming part of our internal narrative of the ever-evolving story of our life.

Formulas and Freedom:
The Wide Range of Experiential Aesthetics

The expansive—and expensive—Scottish Parliament building in Edinburgh, by Enric Miralles and Benedetta Tagliabue (EMBT), embodies all these tenets of experiential design, conclusively demonstrating that following them will create rich, vibrant, and highly varied designs. The Scottish Parliament is nothing if not distinctive, communicating what is often referred to as the "fierce" independence of the people it represents. As large and lavish a building as the St. Nicholas Church is small and modest, the magnificent Scottish Parliament synthesizes these aspects of designing for humans in a built environment that includes urban design, architecture, and landscape architecture. From the point of view of the user, the Scottish Parliament conclusively establishes that the *built environment* is the master category, a trifecta that all designers should master and aspire to design.

This beautifully distinctive building symbolizes and facilitates the Scottish people's democratic aspirations for self-governance; at the same time, it crystallizes the individualistic Scottish national ethos by harmonizing with its surroundings. It is so folded into the streetscape and the landscape that you barely apprehend its immensity. Sited at the lower end of the Royal Mile, across from the British monarchy's Holyrood Palace, the Scottish Parliament viewed from afar echoes the area's mountainous protrusions, appearing practically dug right into the foot of the Salisbury Crags, Edinburgh's volcanic eruption of rocky red cliffs, hiking paths, eddies of water, fields of heather, wispy grasses, and ten-foot-high thistles in shades of bluish gray. From the public entrance facade, grass-covered, stepped embankments hide services and a below-grade multilevel parking garage, curving their way in from the foot of Arthur's Seat and terminating in the zigzaggy reflecting pools that introduce the public entrance. The landscape per-

Scottish Parliament (Enric Miralles/EMBT Architects), Edinburgh, Scotland

forms a directional function similar to the Salk Institute's channel fountain, with spatial and sensorimotor indicators pointing toward and ushering us into the entrance.

Housing the entrance foyer is a long, low barrel-vaulted space top-lit from skylights and dropped atria (a design that, incidentally, recalls another building by Kahn, the Kimbell Art Museum in Fort Worth). Behind this long, low block congregates a series of midrise leaf-shaped blocks containing the assembly chamber and council chambers. The entrance to the debating chamber and the gathering spaces for Scottish Members of Parliament occupy these blocks, which are arranged along a spine like leaves on a branch, with each "leaf" molding chamber spaces that frame views onto the crags, admit and sculpt daylight in distinctive ways, and are exquisitely proportioned to the scale of the human body. The single-story loft-like spine continues the unusual concatenation of indoor and outdoor rooms, as sculptural concrete-and-wood supports open up huge leaf-shaped skylights which break up the horizontal expanse into smaller, individuated spaces. The light-drenched foyer and adjacent, beautifully planted exterior garden graciously offer human-scale indoor and outdoor eddies and seating areas where the parliamentarians can gather among themselves or talk to members of the media.

At the opposite end of this foyer on the city side of the site rises the

Petal chamber, Scottish Parliament

tallest block in the complex, which contains the offices for Members of Parliament and their staff. In a complex of buildings overflowing with multivalencies, the exterior facade of this office block crystallizes patterned complexity's visual appeal and experiential playfulness: body-sized inglenooks for reading or quiet conversation project from the facade, sculpted appliqués pantomiming the easy-chair-plus-stepped-

Exterior office blocks with projecting reading nooks, Scottish Parliament

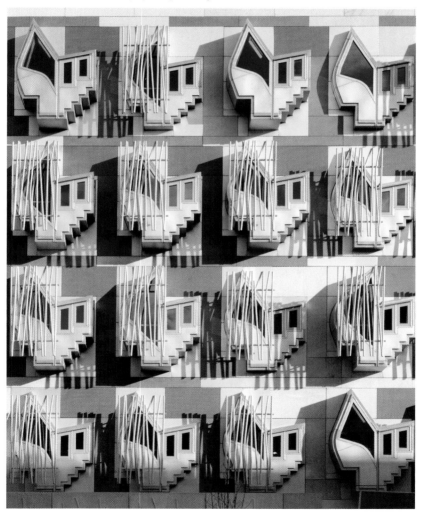

shelves that they house and variable arrangements of sunscreens resemble pickup sticks cast over each individual window.

Inside and out, the office block design celebrates the singularity of each individual within a collectivity, setting the stage for the extraordinary celebration of social congress that we get when we pass back through the petal chamber to ascend the spreading staircase that

Reading nook in each Member of Parliament's office

funnels us into the breathtaking assembly room. This hundred-foot free-span amphitheater, flooded with daylight, admits few views of the monumental landscape nearby, concentrating our focus on the important business at hand of governing a people. Unlike the opposed and confrontational benches of London's Houses of Parliament that Churchill championed, here the seats are arranged in a flattened semicircle, which symbolizes and in actuality allows for greater fluidity and cooperation among Scotland's political parties. Light streams from the windows stacked high up on both walls of the boat-shaped room. The structure permits these openings because the muscular, dynamic ceiling is constructed of spanning beams above, which are shaped to curve toward the midpoint of the load that they carry, along with dropped wooden beams that taper dynamically at their edges. The wooden hull of the ceiling appears to heave slightly under the weight it supports. With deft poetry, the assembly chamber design updates its two sources of inspiration, the timber roofs of grand meeting halls in Scottish castles and the inverted boat hulls that Miralles and Tagliabue, its architects, found beached on the Scottish shores. The assembly chamber promotes a sense of awe not unlike that we experience when visiting the cathedral in Amiens, uplifting our individual spirits and promoting prosocial feelings of our shared, human commonality.

With gentle muscularity, the Scottish Parliament crystallizes the joyfulness of democratic ideals and the gravity of being charged to execute them; at the same time, the complex—an urbane landscape and aggregation of buildings all at once—demonstrates how wide-ranging are the possibilities of experiential design. In no way formulaic, it is the product of how Miralles and Tagliabue, working with partners and clients, poetically interpreted and shaped the identity of a political institution by creating original, varied, and geographically and culturally suitable forms. The design of the building, its landscaping and

Assembly Chamber, Scottish Parliament

urbanism, evince the kind of patterned complexity that slows us down. Over and over we are invited to engage, with all of our senses, to take in its modeled surfaces, its natural light, its visible, muscular construction details, and its richly textured materials, including concrete, Scottish gneiss, granite, oak beams, and thin, light-emitting sheets of sycamore. From its overall design concept to its crafted details, the Scottish Parliament integrates the scale of the human body with the scale of the city and the monumental site. From spatial organization to individual incident, it demonstrates ways that you can live individually and cooperate socially in this highly disaggregated, staggeringly complex world. Themes and metaphors of history, landscape, democracy, individuality, and social cooperation run through the entire design—not just its facade or the exterior envelope—shaping its spaces and giving its users a thorough built environmental experience that will inspire and ennoble many days and lives.

CHAPTER 7

From Blindsight to Insight
Enriching Environments, Improving Lives

Ethics and aesthetics are one.
—LUDWIG WITTGENSTEIN, *Tractatus Logico-Philosophicus*

Building nearly anything consumes considerable resources. Even a modest-sized infrastructural element, edifice, park, or playground must be designed, engineered, financed, permitted—all before construction even begins. Depending upon the size of the project, it's a process that commands the labors of dozens or hundreds of people, at a cost of tens or hundreds of thousands of dollars, or millions—and for the largest projects, even more. Once finished, a new urban area or park or building will likely outlast every person who designed, engineered, and built it. It will survive too the people who wrote and adjudicated the codes that dictated its permitting. And it will remain in use long after those who commissioned and paid for it are gone.

That's why the design of our built environments must not be hijacked by short-term or parochial interests. That's why it must not be unduly shaped by the many other things it so often is: by people's ignorance, by apathy or reflexive antipathy to change, by corruption or greed. In the previous chapter, we examined some design principles that every person who has a hand in the shaping of the built environment should approach as inviolable and nondiscretionary. Here we extend the inquiry to ask: What operational, social, and ethical principles should guide us in the extraordinary task of building the environments of today and tomorrow? Societies and institutions need marquee buildings, landscapes, and cityscapes. Fortunately, sometimes, in some places, they get them. But already we've seen that building well need not mean building lavishly, even if spending heavily on a first-rate design certainly has proved worth the investment in the past and will be that much more valued and valuable in our ever-urbanizing future.

The distinction between building and architecture, between designing for aesthetic pleasure and designing (or building) for "function," is misleading, wrongheaded, and defunct. Everyone needs better—indeed, good—landscapes, cityscapes, and buildings of all kinds, and everywhere. The only way to accomplish this is to continuously take stock of what we know and are learning about the ways the built environment shapes and affects human experience. That involves carefully analyzing how we can use that information to support people's physical health, promote their cognitive development, and foster emotional well-being. An ever-growing body of knowledge is enabling us to more precisely devise appropriate expectations and standards and demand their implementation across the board: by the real estate developers and financiers, the contractors and builders, the code writers and the code enforcers, the city planners and designers, and the users and clients. By all of us.

Nearly anything that anyone helps to construct can and likely will affect many people, and then the lives of generations. So at a minimum, peoples and policy makers should insist—indeed, *require*—that every cityscape, landscape, and building first of all be *designed*, and second, designed by trained professionals, and third, that those professionals be thoroughly schooled in the evolving body of knowledge in environmental aesthetics and experiential design. Practitioners, researchers, and scholars should continue developing programmatic research agendas and supporting studies that will expand our knowledge, and they should use their institutional bases—public entities and nonprofit organizations, think tanks, academic institutions, and areas of the private sector (like residential and health-care architecture)—to conduct such inquiries, and programs to disseminate that information widely should be developed, funded, and promoted. Everyone should work to ensure that experientially enriching design and environmental aesthetics be incorporated into the products that companies manufacture for the building industry; into construction practices and zoning and building codes; and into municipal, regional, and federal review and oversight mechanisms.

Such broad-sweeping changes will come about incrementally, and even then, unevenly. But every step of change is realistic and realizable. One simple example will suffice. In many parts of China, building codes mandate that every newly constructed apartment receive a minimum of three hours of direct sunlight per day on the shortest day of the year, the winter solstice. Think about it. How many people's homes would be improved if just this one law were enacted and enforced across the globe?

The built environment's decision-makers—from real estate developers and corporate entities to private clients, public institutions, and governments—should wholeheartedly embrace the realistic goal of building experientially enriching environments. Unique projects, whether they be landscapes such as New York City's High Line, urban places such as Chicago's *Cloud Gate,* or buildings such as EMBT's Scottish Parliament, attract people to cities and public places. But good design, be it small, medium, large, or extra-large, is an essential factor in promoting human health, development, and well-being. It is good business, improving worker satisfaction and productivity, making retail enticements that much more appealing. It is good social policy, promoting a sense of community, enhancing people's emotional investment in a neighborhood or place. It can even be good politics, because it has the potential to promote civic engagement. Quite simply, good design is the right thing to do.

Promoting and advocating for good design is everyone's responsibility because the built environment, even though largely owned and constructed by private individuals, is akin more to water, energy, or digital communication in the sense that it is fundamentally a social and a *public* good. Yet as we have seen, most polities, societies, and individuals have not treated it as such. The sorry places where most people live fall far short of meeting reasonable standards of living. In most parts of the world, public oversight is minimal at best, and most construction projects are "designed" by people with little to no expertise in the fundamentals of human-centered design. Even projects that are designed by so-called experts can and do violate the tenets of human experience because today, most of the academic institutions that offer professional design degrees do not systematically teach what we know about how people experience, interact with, and are affected by their primary habitats, which are buildings, landscapes, places, cities. Ordinary people as well as the commissioning and decision-making elite cling to the invalidated notion that design is a luxury rather than a necessity, a matter of taste rather than of urgent public welfare. The result? A self-perpetuating system of built environmental beggary.

Society's baseline expectations for design in the places we inhabit are wrong. Maintaining that it is enough for our constructed worlds to hew to utilitarian standards is akin to maintaining that all a person needs for a good life is a little food and water and the assurance that she will be neither poisoned nor crushed. Such benchmarks fly in the face of the vast and growing body of scientific and social scientific knowledge showing that we humans are thoroughly and profoundly embedded—physically, physiologically, and psychologically—in the environments we inhabit and upon which we rely, and that every element—building, landscape, urban area, infrastructure—ought, accordingly, to be designed to help us thrive. With decades of solid research behind us, urbanists, policy makers, developers, designers, and ordinary people must confront and begin to grapple with a truth as troubling as it is inescapable: today's ever-more-rapidly ur-banizing built environments are designed and constructed to standards that are so scandalously low that they are inhumane and unacceptable—unacceptable *by nearly any measure.*

To be sure, not everything about how people experience the built environment cuts across cultural and historical particularities, and in-dividual differences, gender- and age-based differences, and cultural differences exist. (Two quick examples: elderly people perceive the slope of a ramp as steeper than younger ones do, and women seem to gravitate more toward "refuge" spaces than men do.) Still, much about experienc-ing the built environment is something all humans share, because it is a function of how humans adapt to and develop in their environments

Inequality is inscribed into our landscapes: modern apartments adjacent to slum dwellings, Mumbai, India

from infancy into adulthood, imbuing them with meaning, and of how humans collectively have evolved over tens and hundreds of thousands of years living on an earth many millions of years old.

That our concern and primary focus here remains the *contemporary* world changes little of this. Today's trifecta of globalization, mass urbanization, and growing income inequality coupled with unprecedented levels of wealth is directly inscribed into the places we inhabit and build. The grinding poverty of slum dwelling drains people's bodies and spirits. The widespread lack of adequate educational opportunities and the disparities in access to adequate health care are becoming not just morally unacceptable but also economically counterproductive and, increasingly, politically unfeasible.

The transitory, intercontinental flows of people, goods, and information has heightened many people's appreciation of the fecund heterogeneity of human cultures and societies, but at the same time they have exacerbated people's sense of disconnectedness from place. The revolution in digital technology confounds our innate desire to maintain control over our surroundings. Globalization has ignited fears— sometimes well founded—that disintegrating traditions will leave only social and cultural rootlessness in its wake. Shopping areas, airports, indeed whole new cities and suburbs, seem placeless, as though they could be anywhere. In the academy, a whole new literature decrying the pervasiveness of "non-place" places, has burgeoned. These realities already have changed, and will continue to transform how we conceive of ourselves and our place in the social and built worlds. Climate change is also forcing peoples and countries to build differently and use resources and sites with more thoughtful deliberation, compelling us to rethink the built world's relationship to the natural world in toto.

All this can and should continue and improve in the coming decades. We have a colossal amount of building to do. The status quo must not continue. Today, every day, children around the world, especially

"Non-place" places: cityscape, China

underprivileged ones, are robbed of opportunities for social advance-
ment and self-actualization. One large part of the reason why is that
they live in unhealthy or cognitively dulling habitats and attend school
in buildings that put a drag on or literally undermine attention, motivation, and effective learning. Every day, millions of people fail to find
comfortable, inviting, well-designed streetscapes, buildings, parks, and
plazas where they can simply escape the stresses of daily life or share
the easy company of others. Every day, the least privileged members of
our society return to inhospitable, decrepit, soul-deadening homes, including the many hideously ugly "affordable housing" developments in
my own neighborhood, East Harlem. The world is literally littered with
places that were built on the cheap and violate practically everything
we now know about what makes for salutary, enriching environments.
Such places take their substantial toll, as they practically shout in the
faces of their occupants every day that *their* lives don't matter—at least,
not to the people who govern and shape the society and polity of which

they too are members. Today, well into the twenty-first century, all this is much more scandalous because *now we know* that design matters in people's lives, and that it matters in lasting, profound, and indeed, in foundational ways.

Enhancing Human Capabilities

People usually discuss and evaluate design in the narrowest of terms: is this "good" or "bad," tasteful or crass, practical or extravagant. Overly simple and dichotomized criteria mean little, and we must begin to evaluate our built environments from a different, larger framework. A good starting point comes from the well-established position of Martha Nussbaum, a philosopher, and Amartya Sen, an economist, who argue that the obligations of a polity to its citizens extend well beyond the provision of political stability and economic sustenance. A well-ordered, ethically justifiable society, maintain Nussbaum and Sen, must guarantee more

Some housing shouts in the face of occupants that their lives don't matter: East Harlem, New York

than *negative freedoms,* by which they mean our right to be free from political and social institutions that thwart the pursuit of basic human needs, such as to protect, feed, and educate oneself and one's family. The well-ordered, ethically just society ought to actively support and promote *positive freedoms,* by which they mean the liberty to develop our individual capabilities so that each one of us has the tools and positions to pursue a full, successful, and meaningful life. From Nussbaum and Sen's orienting question, "What is each person able to do and be?" they propose a roster of political, social, and cultural conditions that would establish the fullest, correct meaning of *human development,* whereby every person inhabits a society that allows him or her to develop his or her particular capabilities of body and mind, of individual spirit and social connections.

Nussbaum and Sen's Capability Approach is neither effete academese nor utopian fantasy. It has centrally influenced contemporary international human rights thinking and policy initiatives. The World Bank and the United Nations Development Programme have adopted its central premises and endeavored to operationalize them. Indeed, Nussbaum and Sen's Capability Approach underlies the UN's annually updated Human Development Index (HDI), a widely used measure of the health of countries and peoples around the globe. So extending these precepts to the standards by which we evaluate today's and tomorrow's built environments makes good sense.

What is each individual able to do and be? Nussbaum and Sen explain that the processes by which we cultivate our individual capabilities and become productive members of society rely on a roster of conditions that only properly oriented political and social institutions can provide. Inferring the built environmental dimensions of them is easy. Each individual must be guaranteed both physical safety and some minimal standard of bodily and mental health. They should be able to reasonably anticipate the availability of both, for themselves and their families, in the present and in the foreseeable future. A good education inculcates children with the foundations for the lifelong project of self-actualization; this includes imparting practical knowledge of the social norms necessary to participate actively in society, and the critical thinking capacities at the root of practical reasoning and supple, creative imagining. In addition, people should be positioned to "live with and toward others" (Nussbaum's words) through the emotional connections they develop to groups and the institutions in which they and their social reference groups are embedded.

Good design—landscape, urban, and architectural—is an essential factor in all of these. Consider housing, the obvious example because it figures in the guarantee of safety, and physical and mental health. Properly designed housing can and must vary by economy, region, culture, and individual needs. Two government-sponsored projects in South America help to illustrate the wide range of possibilities. For Iquique, a port city in northern Chile, Alejandro Aravena's firm Elemental grabbed the paltry budget on hand to design and construct low-cost housing, devising a multifamily development of what they called "half a house" homes. Three-story townhouses, upon completion, supply just enough space to house a family in toto. But the firm also included walled-in empty areas that families could use as a patio or, if the owners' fortunes improved, to inexpensively construct an extra room. The Quinta Monroy project was the first of such projects, which since have been constructed in Santiago, Chile, and in Monterrey, Mexico.

Also in Mexico, where the housing crisis is so severe that an estimated 9 million people lack homes, the architect Tatiana Bilbao proposed a related but different approach. After interviewing the potential residents of a government-sponsored low-income housing project, Bilbao discovered that for them, a crucial source of dignity lay in three trappings of a traditional middle-class home: the sense (if not the actual experience of) spaciousness, pitched roofs, and the *appearance* of com-

Positive freedoms: "half a house" to start: Quinta Monroy Housing (Alajandro Aravena/
ELEMENTAL), Iquique, Chile

pletion. (In Mexico, as in many other parts of the developing world, people use reinforced concrete to construct as much of a house as they can
afford, and leave the metal rebar exposed, sometimes for many years,
as they try to earn the money to build that hoped-for second or third
story.) With the information she gleaned from interviews, Bilbao constructed for only $8,000 a solid core house that is more spacious than
the state-imposed 460-foot minimum. Costs were kept down by using
inexpensive and sometimes recycled materials such as concrete block,
plywood, and wood pallets. Bilbao's prototype single-family residence,
the Sustainable House, sports the pitched roof residents desired, and
contains a living room that seems unusually spacious owing to its high
ceilings, which also promote wind circulation and help with ventilation
and climate control. The Sustainable Housing looks finished. Yet because it is based on a modular system, families have many possibilities

for how they can inexpensively and incrementally expand the house, adding an extra room here, a mud room or partially enclosed outdoor space there, and so on.

Such projects—and there are many other impressive examples all over the world, such as Vo Trong Ngia's S House, discussed in the previous chapter—illustrate how design, and even the design of low-cost housing, could and should be a component of Nussbaum and Sen's Capability Approach. However, because the information demonstrating the centrality of the built environment's design to people's lives is not widely known, Nussbaum finds herself at a loss about what to say when it comes to discussing its role in the development of human capabilities. Her pathbreaking book *Creating Capabilities,* the fullest account of her and Sen's paradigm, contains scant commentary—really, next to nothing—about either architecture or landscape architecture or urban design. Nussbaum clearly expresses her appreciation for their impor-

Modular sustainable housing (Tatiana Bilbao Studio), Chiapas, Mexico

tance. Yet she writes only that "having decent, ample housing may be enough . . . the whole issue needs further investigation." Such uncharacteristic brevity can be understood as an unfortunate artifact of the built environment's pervasive marginalization and neglect, rather than as an indicator of its objective importance, and certainly not of its relevance to the development of human capabilities.

When the design of our domiciles and institutions more reliably accords with the principles of human experience, they nurture and sustain the very capabilities Nussbaum, Sen, and many other thinkers and policy makers on the subject of human development champion. It is established fact that children develop better in spacious, sturdy, quiet, orderly homes. That they learn much more effectively in a well-designed school than in a poorly composed one. Bodily health is best nurtured in facilities designed with the most up-to-date information on environmental psychology and cognition. Experientially designed places of leisure substantively mitigate stress and help to restore easily depleted attentional faculties, and foster creativity. Workplaces can be configured to enhance people's problem-solving, interactive skills, creativity, and focus. Design techniques and decisions of all kinds can encourage the kind of prosocial conduct that strengthens communities. In terms of social betterment—indeed, in terms of social justice—all this adds up to a claim that is at once bold and obvious: the commonly used instruments of public policy, private investment, and philanthropy such as health care, infrastructure, teacher training, and primary and secondary education would be far more effective than they are today if the built environments in which they more generally took place were better designed.

Human Capabilities and Enriched Environments

Built environments that accord with the fundamentals of human experience constitute what we should call "enriched environments," at once a self-explanatory term and one that carries specific meaning for the small community of scientists who study the relationship between environments and cognition. Picture a rat in a standard-sized cage, outfitted with a spinning wheel: call this control setting an impoverished environment. Now compare that to a second cage. Same rat. The container is a little larger, though, because alongside the running wheel are arrayed other playground-like toys: a little slide, perhaps; a pool, a ladder, a maze. From our rodent-sized vantage point, the second cage presents plentiful and engaging things to do, places to hide, hurdles to jump, pedestals to survey the surroundings or to preen. It is, in other words, an enriched environment.

Rats that live in enriched environments, compared to ones in the running-wheel-only environments, flourish. They are more resilient to stress. They display superior skills of spatial navigation. Their visual systems function better and are better coordinated with their motor systems. They learn (and preserve long-term memories) more easily, and their brains are better defended against the cognitive decline that comes with age. Humans are unlike rodents in countless ways, to be sure. But we are like them in this: we too are permanently diminished when exposed primarily to impoverished environments. And when we can enjoy the manifold boons and opportunities of enriched environments, we flourish.

Of course well-designed, enriched environments nurture human capabilities! Throughout this book, we have encountered examples of such experientially designed urban areas, buildings, and landscapes. Many, many more such places exist, all over the world, fortifying our confidence in their relevance and durability: places such as the Luxembourg Gardens in Paris and the Ssamziegil mall in Seoul and the 798 Art District in Bejing. Some encourage us to restore our fragile attentional resources by allowing us to freely explore and to creatively imagine. Some, like Amiens Cathedral, Soufflot's Panthéon in Paris,

Enriched environments: 798 Art District, Beijing, China

and the Sydney Opera House, awe us into a sense of our commonality with others and humility in the face of the natural world's mysteries. Others, like the National Pensions Institute in Helsinki or the Herman Miller plant in Michigan, the Bukchon teahouse in Seoul or the Sidwell Friends School in Washington, DC, create settings optimal for concentrated work by helping us to focus our attention on the tasks at hand. Still others—Chicago's *Cloud Gate* and Millennium Park surrounding it, Antwerp's Museum at the Stream, La Jolla's Salk Institute—captivate by stopping us short, forcing us to problem-solve by challenging our assumptions and expectations.

They and other projects engage us—seduce us, even—by drawing us into a whole-body, multisensory, and cognitive engagement with them. They create the circumstances for us to become smarter, more resilient, more flexible problem-solvers. Their overall forms, materials, and details are composed to accord with and build on the associative, nonconscious ways that we, as humans, experience the world.

Imbued with character, they are layered with meaning and interlaced with primes, embodied and situated schemas and metaphors. Deliberately constructed as action settings that express the nature of the social institutions they house, they enrich our experience of them as places and as physical instantiations of our communal lives. And because of the memories we form in such settings and then recall and rely upon for the rest of our lives, they literally create the framework whereby we define and conceive of who we are. Enriched environments, whether attention-focusing or attention-restoring, whether awe-inspiring, defamiliarizing, or plainly comforting, will always be the habitats best suited to the human project of personal, familial, and communal well-being, self-actualization, and accomplishment.

Understanding this changes the nature of our political and social responsibility for the built environment. It establishes incontrovertibly that experiential design is not optional. Life-enhancing opportunities literally can be built into the enduring land- and cityscapes of our lives—our homes, our children's schools, our workplaces, our streets, our parks, urban areas, and playgrounds. As we establish the design principles that enriched environments entail, they should be incorporated into the core of the Capability Approach, and be incorporated into development indices worldwide, including the World Happiness Report and the UN's Human Development Index. Human-centered, experiential design should be seen as the fundamental human right that it is.

Enriched Environments Foster Conscious Cognitions; Conscious Cognitions Promote Agency

One last important dimension of enriched built environments remains to be discussed: their capacity to shift us out of our ordinary, nonconscious, egocentric point of view. If we conceptualize human consciousness as a spectrum from nonconscious to conscious cognitions, then an enriched environment can slide us across that scale, toward a more conscious state of awareness. Antonio Damasio writes that humans and animals both "form intentions, formulate goals, perform actions" in the process of collecting information from the environments they inhabit, but as far as we know, only people "have the capacity to do these things while at the same time using the internalized schemas of the space of their bodies and the space around their bodies to contemplate what they are doing, why they are doing it, and where they are, or are not." This ability to think and act while at the same time watching ourselves think and act is called metacognition, an awareness that rests in part on our human capacity to conceptualize ourselves in the third as well as in the first person; or if we shift from literary terminology to environmental imagery, allocentrically, from a point of view outside ourselves, as well as egocentrically, from the inside, from the zero degree of our own body. Unlike the nonconscious and distracted ways in which we mostly apprehend our environmental surroundings, during conscious thought, we experience ourselves as physical, thinking beings from a hypothetical allocentric perspective, imagining ourselves as bodies in space, among other people and other objects.

For a design to nudge—in some cases, to shove—our cognitions over from the nonconscious end of the spectrum toward the more or wholly conscious end, the designer must devise ways to make us shift out of our ordinary, habit-driven state and *attend* to our surroundings. The environment itself makes us conscious of the interactive nature of our relationship to our own bodies, to the natural world, to the social world, and to itself, and only then can we reflect upon our experience from multiple points of view. Why is this desirable? Because promoting the awareness that we ourselves are discrete, situated beings in a particular place, at only this and no other moment in time, fosters our sense of ourselves as both individual agents and collective actors in our worlds.

What is each person able to do and be? No matter where we reside, we need the sense that we play a role in the shaping of our days and weeks, that we exercise some measure of control over the trajectory of our lives and the ways we live them. By fostering memorable experiences, enriched environments enhance our sense of place. And they lay the groundwork for us to take more of an active role in the shaping of our constructed worlds. Each new enriched environment could help effectuate the start of a momentum that will reverse the cycle of self-perpetuating beggary in which most people live, helping to put in motion and propel a self-perpetuating cycle of virtue in its place. People will expect and actively demand more from the places they inhabit. They will work harder to ensure that what gets built meets these higher expectations.

Looking and Moving Forward

More than ever, today we are all in a position to expect, insist upon, and help to create more enriched environments. We understand many orders of magnitude more about people's experiential needs, and with each passing year, that knowledge grows. Climate change has heightened people's general awareness about the interdependence of environments, built and natural, instigating a worldwide discussion about how we manage the earth's resources and landscapes, which will be ongoing for decades to come. Already, one felicitous consequence of climate change is that it has changed the way designers operate. As we saw in the Scottish Parliament, which was finished in 2004, the three professions—architecture, landscape architecture, and urban design—are starting to reintegrate and collaborate after decades of operating in more or less distinct silos. New digital technologies of design and manufacturing have the potential to make it even more possible and cost-effective for designers to address an ever-wider range of people's experiential needs, even at the mega-scale that today's cities demand. Computer-aided design offers a powerful array of tools that greatly expand the range of forms that can be cost-effectively engineered and manufactured or constructed. Digital modeling enables practitioners to explore countless formal iterations of unprecedented complexity while maintaining a design that adheres to prespecified parameters. Inexpensive 3-D printing now enables designers to quickly model ideas. More and more tools will be available to run performance and other kinds of tests on these models. Computer simulations of auditory, wind, and other conditions have existed for years. Virtual reality has so greatly improved that neuroscientists can now study people's neurological and psychological reactions to simulated environments in real time, as they currently do with the StarCAVE at the University of California, San Diego.

Advances have been made in the field of construction as well. The possibilities of computer-aided manufacturing, through computer numerically controlled production processes—analogous to three-dimensional digital printing, but for entire building components—are ever-expanding. As a result, designers are, for the first time in the built environment's history, able to uncouple the mass production of building elements from the curse of simplistic repetition. The Aqua Tower by Jeanne Gang of Studio Gang in Chicago, which was built by a real estate developer admirably ready to try something new, stands as just one impressive example of how much simpler and more economical it now is to incorporate irregular and curved forms into large-scale designs. If curving surfaces entice us, architects can now design them more frequently than before. Digital computation made it economical to manufacture Aqua Tower's swerves and irregular floor slabs on site. The lilting facade serves multiple purposes both "functional" and "aesthetic." They stabilize the high-rise against wind forces by "confusing" the winds' paths. Balconies are shaped to correspond to the unit's internal layout, views, and orientation to the sun. When Aqua Tower's swerves are seen from the proper vantage point (which includes the distance-view from which most Chicagoans see the building), they impart a kind of op art effect. A static structure, paradoxically, appears to move, to ripple, almost, in Chicago's legendary winds.

The technological innovations underlying the design and construction of Aqua Tower make it easier for designers to produce projects that will sustain people's attentions and interest over time. Surfaces and compositions can be made to seem actively, even restlessly figural from every angle, even though they are in actuality inert. Nader Tehrani of the Boston-based firm NADAAA, working with John Wardle Architects of Melbourne, did this in the School of Design at the University of Melbourne (MSD), a building that gracefully integrates all the climatological, material, and social circumstances of its use with the building's pragmatic and mechanical functions, and brims with stop-short moments and innovative ideas. The architects inflected each part of this large academic building to the conditions of its site, so it serves simultaneously as an active pass-through from one part of campus to the other and as a social incubator for the school's design community.

Spanning its seventy-foot-wide central atrium is a wooden beam structure in laminated veneer lumber, modeled into deep, irregularly shaped coffers. Each coffer's angles are canted to most effectively diffuse the bright Australian light streaming in from above, creating an origami-like structure that houses and hides the systems for ventilation and lighting. Near one end of the atrium, this heavily modeled structure dramatically erupts from the ceiling into a hung, perforated sculptural form that offers views into its habitable spaces; it houses interior classroom and studio presentation spaces.

Aqua Building (Studio Gang), Chicago, Illinois

The Melbourne School of Design interior stops us short because it deploys and challenges our form-identification, pattern-detection, and trajectory-completion schemas. Like the Aqua Tower facade, but with even greater skill and complexity, MSD entices us perceptually—over and over again—into perceiving motion when none exists. We cannot just *look*. We cannot but move around and explore this unusual space, trying to work out the profiles of their constructions and the contours of their forms. MSD also exhibits how fractalization can help to establish and reinforce a human sense of scale while generating patterned complexity in its overall form, the deployment of its materials, and the nature and finish of its surfaces.

Aqua Tower and MSD only begin to demonstrate the promise that digital technologies hold out for designers who wish to explore the many affordable avenues that contemporary designers can explore in order to better address human experiential needs. Today more than ever before, with age-old materials and means as well as new ones, any place can be inflected to the complexities of our various social worlds, and to the specificities of the site, the body and its multisensory systems, and our experience through the passage of time.

We have much to do, and on a vast scale. But let's not lose sight of the good that every improvement will bring. Big changes can start with small improvements, to this apartment building, that house, this facade, the neighborhood community center, the big-box store, the neighborhood playground or park, the urban square, that office or civic or cultural building. The built environment, like a society of individual people, is a society of individual buildings, structures, places, and landscapes. Each one of them can be enriching or soul-deadening by design.

If you think it's unrealistic to set in motion a virtuous cycle of good design, consider the Netherlands. The baseline quality of the Dutch built environment outclasses the United States by nearly every measure—design and aesthetics, quality of materials, and quality of

School of Design (NADAAA with John Wardle), University of Melbourne, Melbourne, Australia

Playing at Geopark: Geopark (Helen & Hard), Stavanger, Norway

construction. This is true of marquee projects and vernacular ones, in big cities and in small towns. Why? First, design education takes place within an intellectual context that welcomes new knowledge from fields like environmental psychology and health care. Second, most municipalities require that new projects undergo aesthetic review by a committee of licensed design professionals, whose membership rotates regularly. Third, because of all this, the Dutch are habituated to better design: they live with it, they expect it, and they demand it. As a result, standards for materials and construction, even for ordinary buildings, are higher. This is not to say that Dutch design represents the apex of development in good design; it does not. But the Dutch have grown accustomed to better buildings, and so they get them.

For good and for ill, buildings and cityscapes and landscapes literally shape and help constitute our lives and ourselves. Designing and building enriched environments, ones that are informed by what we now know and are learning about how people experience the places they inhabit, will promote the development of human capabilities. Just as is true with regard to global warming and the earth's environment, nearly everything we construct today will outlast us to affect those who come after us, sometimes generations and generations of them. Shouldn't a better built environment be the legacy we leave to the world?

Acknowledgments
and a note on photography

The genesis of this book was threefold.

Shortly after we'd met, my husband, Danny, gave me George Lakoff and Mark Johnson's little book entitled *Metaphors We Live By*. In it I found a powerful way of thinking about thinking and human experience which was and remains utterly at odds with the postmodern paradigm of human cognition, which continues to be espoused by many humanists in the academy. While postmodern theorists maintain that human thought and experience is wholly socially constructed, Lakoff and Johnson outlined a different approach, which is called embodied cognition and is grounded in the principles of human psychology. This resonated with and clarified ideas I'd been struggling to articulate on my own for some time.

A few years later, at Stanford Anderson's invitation, I wrote an essay on the great modernist architect Alvar Aalto's use of metaphors, taking off from Lakoff and Johnson's ideas after having discovered that Aalto had immersed himself in scientific studies of human perception and psychology. Stanford and his coeditors, David Fixler and Gail Fenske, waited patiently for me to write that essay, then published it in their book *Aalto and America*. Stan passed away in January 2016. Here is the book I am sad not to be able to hand to him directly with an expression of my gratitude for his early support for these ideas.

While I was writing the Aalto essay, I received an unsolicited email from someone I'd never heard of, or met: Chris Parris-Lamb, now my agent. Explaining that he'd read and liked my criticism for the *New Republic* as well as my first book, Louis Kahn's *Situated Modernism,* he was wondering: Had I ever considered writing a book for a general audience? By then I'd realized that in embodied cognition, along with the psychology and science of mind, I'd found a way to make the strongest possible case for why the design of the built environment matters a lot more than most people really realized it did. Within seconds of reading Chris's email I hit the "Reply" button and wrote, Yes. Let's talk.

That was in 2009. To make the case about how critical the design of the built environment is to people's lives, I was going to need to delve into areas of scientific and psychological inquiry with which I'd had only glancing experience until then. During the subsequent years of research and writing, I relied on the kindness and company of strangers and friends—and of strangers who became friends. I am indebted to those who read and responded to drafts along the way. These include architects and people allied with architecture—Louise Braverman, Rosalee Genevro, Nader Tehrani. They include people whose work also deals with embodied cognition and the scientific study of human cognition: Harry Francis Mallgrave and Barbara Tversky. And they include people dear to me, whose intelligence and trenchant insights I seek out for many things, and whose friendship I treasure: Mikyoung Kim, Lizzie Leiman Kraiem, my brother, Roger S. Williams, and my sister, Joan Williams.

Researching and writing this book—for which I had found no adequate models—felt, at times, like lonely, difficult work. At moments, I grew discouraged. These friends and my family, along with Matthew Leeds, Upali Nanda, Peter MacKeith, and Terrence Senjowski, expressed support for what I was doing at critical junctures during this long process, for which I was then and remain grateful. David van der Leer and Anne Guiney at the Van Alen Institute enthusiastically endorsed my ideas. Matthew Allen invited me to present my ideas at the conference he organized, "Architecture, Beginning with the Brain," which was held at the School of Architecture at the University of Toronto. It was helpful to have the opportunity to meet like-minded people and test out some ideas in real time, as it was again when I presented the following year at a conference sponsored by the Academy for Architecture and Neuroscience at the Salk Institute in La Jolla.

Many scholars, scientists, writers willingly shared their work and ideas with me, most notably Terrence J. Senjowski, Linda B. Smith, Joseph Biederman, Stephen Kellert, Brook Muller, Barbara Tversky, and Stephen and Rachel Kaplan. Robert Condia and Kevin Rooney at the University of Kansas generously shared with me their research and ideas. Daniel Lende, a neuroanthropologist from the University of South Florida, engaged in productive correspondence with me after I released a trial balloon into the atmosphere in the form of an op-ed for the *New York Times*. And Charles Montgomery, author of *The Happy City*, simply forwarded to me a stack of his own research that he'd gathered in preparation for his book, after I met him at the opening of an exhibition in New York organized by the BMW Guggenheim Lab. Having come from a part of the academy where, among strangers, naked competition, not fuzzy collegiality was the norm, Montgomery's was a gesture of generosity that I shall not forget.

Illustrating the book required a hefty sum of money. Jonathan Burnham and Gail Winston, my publisher and editor at HarperCollins, graciously stepped up to the plate. So did the Graham Foundation for Advanced Studies in the Fine Arts with a grant that helped pay for the photographs and permissions.

Every published author, I imagine, leans heavily on the team of people standing in her corner; in mine, I am fortunate to say, stand some of the best. Chris Parris-Lamb, my agent, and Gail Winston, my editor, responded to drafts of chapters and my dozens of queries on matters related to the book's content and production with a salutary admixture of sympathy, intelligence, hard-headedness, and cheerleading. Toby Greenberg found most of the 170-plus photographs, negotiated fees, and navigated the labyrinthine and utterly opaque (to me) world of permissions and reproduction rights. She dealt with my innocence about these matters as well as my occasional expressions of impatience with grace and good humor. Once the book was in production, Sofia Groopman at HarperCollins managed that process with infectious good cheer.

One person deserves the greatest share of my gratitude: my beloved husband, Danny, with whom these acknowledgements rightly begin, and end. Danny's analytical brilliance, his wit, and his fierce, determined belief in the rightness of this inquiry and project is laced through every paragraph, every concept, and every chapter. Most of the ideas here were first the subject of conversations between us while looking at buildings and touring cities, during walks in biophilic settings, over lunch, or sitting in our living room, first in Newton, Massachusetts, then in East Harlem. Most of us have been witness to bad marriages and good marriages, and some of us have been unlucky and lucky enough to have experienced both kinds. To have my husband be my dearest friend and my toughest, most incisive interlocutor, and my staunchest supporter is to have good fortune indeed, and I may be forgiven for thinking that our love is the greatest love, a love like no other. *Welcome to Your World* is dedicated to Daniel Goldhagen, with our shared hopes for a better and more just future for our children and all children. That's the project to which each of us, in our different ways, has devoted our professional lives.

And a note on photography

In Chapter 1 I describe at length how photography distorts a correct understanding of buildings, cities, and places—yet, for obvious reasons, this book is filled with pictures, mostly taken by others but occasionally by me. I urge the reader not to take these pictures as stand-ins for the real thing. They are provided merely as aids to understanding. Several principles dictated my choice of images, and we managed to abide by those principles in all but a handful of cases. These are: No shots taken from places where a person couldn't or wouldn't actually stand. No night photographs, because these usually are shot when a building is gussied up for the big night out, which isn't often. And no architect's renderings, because digital technology's sophistication is such that an image can look convincingly real while bearing scant relation to how the place it depicts is experienced by people with their feet on the ground.

Graham
Foundation

Published with support from the Graham Foundation for Advanced Studies in the Fine Arts.

Image Credits

Page xvii, top: Construction site, United States. *Zoonar GmbH/Alamy Stock Photo.*

Page xvii, bottom: Construction site, China. *Zhang Peng/LightRocket via Getty Images.*

Page xix, top: Site of New Ordos City, China, before (with artist Ai Weiwei).

Page xix, bottom: Ordos, China. *Ashley Cooper/Specialist Stock RM/AGE Fotostock.*

Page xxvi: The House of Commons after the London Blitz. © *Mirrorpix/ Bridgeman Images.*

Page xxviii: Paths, edges, nodes, landmarks: the central tenets of wayfinding, from Kevin Lynch's *Image of the City.*

Page xxxiii: "There's something about a 150-foot ceiling that makes a man a different kind of man," said Louis Kahn about the Baths of Caracalla (reconstruction). *The Print Collector/Alamy Stock Photo.*

Page 3: Life on the Route des Rails one year after the catastrophic 2010 earthquake, Port-au-Prince, Haiti. *Ruth Fremson/The New York Times.*

Page 5: Slum dwelling in Africa. *iStock.com/Delpixart.*

Page 7: Middle-class developer-built suburban home. *iStock.com/ Paulbr.*

Page 8: Higher-end developer-built suburban home. *iStock.com/Purdue9394.*

Page 11: A school building that helps students learn: Sidwell Friends Middle School (Kieran Timberlake), Washington, DC. © *Albert Vecerka/Esto.*

Page 13: 2010 Serpentine Pavilion (Jean Nouvel), London, England [demolished] © *James Newton/VIEW.*

Page 15, top: Seed Cathedral, 2010 World Expo Pavilion (Thomas Heather-wick), Shanghai, China [demolished]. *Photo: Iwan Baan.*

Page 15, bottom: Seed Cathedral, detail showing seedpods encased in resin rods. *Kevin Lee/Bloomberg via Getty Images.*

Page 18: Faulty design can kill: in 2007, the I-35W bridge in Minneapolis collapsed during rush hour, killing 13 and injuring 145 people. *AP Photo/ Morry Gash.*

Page 21: Harmful noise levels on New York City subway platforms. *AP Photo/ Mary Altaffer.*

Page 23: Stress, noise, and crowding in developing countries, Dhaka, Bangladesh. *AP Photo/John Moore.*

Page 26, top: Highway commuting. *Scott Olson/Getty Images.*

Page 26, bottom: "Boxed, bleached sameness": Las Vegas suburbs, aerial view. *iofoto/Shutterstock.com.*

Page 28: Landscape design in most suburban developments barely exists. *IP Galanternik D.U./Getty Images.*

Page 29: More planted than designed: Rose Fitzgerald Kennedy Greenway, Boston. *AP Photo/Elise Amendola.*

Page 33: The architect's rendering of One WTC tower (at left, Skidmore, Owings & Merrill), depicts a light base and an angled shaft, neither of which was executed as designed. New York, *SPI, dbox via Getty Images.*

Page 37: Photographs distort and prettify: Secretariat and Chamber of Deputies (Oscar Niemeyer), Brasilia, Brazil. *Magda Biernat/OTTO.*

CHAPTER 2: BLINDSIGHT

Page 49: Thinking inside the box. *iStock.com/4x6.*

Page 51: Denver Art Museum (Daniel Libeskind). *Photo by View Pictures/UIG via Getty Images.*

Page 53: Streetscape, West Village, New York City. *JLP Photography.*

Page 56, Opposite: Wood-handled banister in a bright yellow stairwell, Paimio Sanatorium (Alvar Aalto), Paimio, Finland. Photo: *Maija Holma, Alvar Aalto Museum. 2014.*

Page 57, top: Stairs with forced perspective, Palazzo Spada (Francesco Borromini), Rome, Italy. *Cultura RM Exclusive/Philip Lee Harvey/Getty Images.*

Page 57, bottom: Just looking at an excessively rough texture can provoke thoughts of retreat: Yale Art and Architecture Building (Paul Rudolph), New Haven. *Peter Aaron/OTTO.*

Page 60: Body position influences mood: forcing your lips into a smile creates a "smiling" feeling. *Image Source/Getty Images.*

Page 63: Grids, Jean-Nicolas-Louis Durand, *Partie graphique* (1821).

Page 65, Above: Grids in houses: Weissenhofsiedlung, House 16 (Walter Gropius), Stuttgart, Germany. *Bauhaus-Archiv, Berlin.*

Page 65, Above and Left: Grids facilitate construction: Weissenhofsiedlung, House 17 (Walter Gropius), Stuttgart. *Bauhaus-Archiv, Berlin,/© 2016 Artists Rights Society (ARS), New York/VG Bild-Kunst, Bonn.*

Page 65, Below: Gridded Cities: Ludwig Hilberseimer, from *The New City. The Art Institute of Chicago, IL, USA/Gift of George E. Danforth/Bridgeman Images.*

Page 66: Grids do and don't fit people: *Images of Life* (Superstudio) © *CNAC/ MNAM/Dist. RMN-Grand Palais/Art Resource, NY (*

Page 67: Wright saw "things out of the corner of his eye": Hanna House (Frank Lloyd Wright), interior. *Photo by © Ezra Stoller/Esto © 2016 Frank Lloyd Wright Foundation, Scottsdale, AZ/Artists Rights Society (ARS), NY.*

Page 68: Triangulating the location of our bodies with two points (or objects) in space, our brains use grids made not from squares or rectangles, but from triangles and hexagons, to help us navigate through space: Hanna House, plan. *John Brandies; redrawn under supervision of Robert McCarter.*

Page 71, top: Retail interior in red. © *Albert Vecerka/Esto.*

Page 71, bottom: Retail interior (Prada) in pale green. *Rainer Binder/ullstein bild via Getty Images.*

Page 73: Lasker Pool, Central Park, New York City. *Sarah Williams Goldhagen.*

Page 74, top: London Aquatics Centre (Zaha Hadid), London, England. © *Hufton + Crow/VIEW.*

Page 74, bottom: Water Cube (National Aquatics Center, PTW Architects with Arup Engineering), Beijing, China. *Photo: Iwan Baan.*

Page 78: Trenton Bath House (Louis Kahn and Anne Griswold Tyng), Trenton, New Jersey. *Marshall D. Meyers Collection, the Architectural Archives, University of Pennsylvania.*

Page 80, top: Aerial view of the Holocaust Memorial (Peter Eisenman), Berlin, Germany. *ZEITORT/ullstein bild via Getty Images.*

Page 80, bottom: People relaxing at the Holocaust Memorial, Berlin. *Bork/ullstein bild via Getty Images.*

Page 86: Autobiographical memories come packaged by place: 74 Allison Road, Princeton, New Jersey. *Sarah Williams Goldhagen.*

CHAPTER 3: THE BODILY BASIS OF COGNITION

Page 93: How others see our bodies—allocentrically: Masaccio, *Expulsion from the Garden of Eden*, 1425. *Erich Lessing/Art Resource, NY.*

Page 94, left: Sensory homunculus. © *The Trustees of the Natural History Museum, London.*

Page 94, right: Motor homunculus. © *The Trustees of the Natural History Museum, London.*

Page 97: The dimensions of a traditional Japanese tatami mat fit the human body. © *Ezra Stoller / Esto.*

Page 98: Designed for many ages and modes of experiencing: Geopark (Helen & Hard), Stavanger, Norway.

Page 99: "Mechanics of Sitting: Dining, Reading, Relaxing, Dressing." Different activities require different body positions, *Aalto: Architecture and Furniture* (Museum of Modern Art).

Page 100: Bent tubular metal looks and feels cold: Wassily Chair (Marcel Breuer), 1926. *Digital Image © The Museum of Modern Art/Licensed by SCALA/Art Resource, NY.*

Page 101: Showcasing the difference between the path of the feet and the eyes: Poly Grand Theater (Tadao Ando), Shanghai, China. *VCG/VCG via Getty Images.*

Page 103: Designing for seeing, hearing, touching, and moving: St. Benedict Chapel (Peter Zumthor), Sumvitg, Switzerland. *Nicholas Kane/arcaidimages.com.*

Page 105: Designing for looking, imagining, and feeling: Notre-Dame du Haut (Le Corbusier), Ronchamp, France. *akg-images/Schütze/Rodemann.*

Page 107: *Cloud Gate* ("The Bean," Anish Kapoor), Millennium Park, Chicago. *Charles Cook/Lonely Planet Images/Getty Images.*

Page 112: Stairs are made for walking, Itamaraty Palace (Oscar Niemeyer), Brasilia, Brazil. *akg-images/picture-alliance.*

Page 115: Glass Pavilion, Toledo Museum of Art (Kazuyo Sejima/SANAA), Toledo, Ohio. *Photo: Iwan Baan.*

Page 116: Proprioceptive challenges: Bioscleave House (Arakawa), Long Island, New York. *Eric Striffler/The New York Times/Redux.*

Page 118: Museum at the Stream (Neutelings Riedijk), Antwerp, Belgium. © *Paul Raftery/AGE Fotostock/VIEW.*

Page 119: Window-curtains: Museum at the Stream, Antwerp. © *Paul Raftery/ AGE Fotostock/VIEW.*

Page 122: Hand-dressed sandstone with hand-bolts on exterior, Museum at the Stream, Antwerp. © *Paul Raftery/AGE Fotostock/VIEW.*

Page 125: Cathedral of Notre-Dame, Amiens, France. © *Javier Gil/AGE Fotostock/VIEW.*

Page 128, Above: Portal, Amiens Cathedral. © *Bruce Bi/AGE Fotostock/VIEW.*

Page 128, Left: Interior, looking up at transept crossing, Amiens Cathedral. © *Vanni Archive/Art Resource, NY.*

Page 129, Opposite: looking down nave, Amiens Cathedral. *McCoy Wynne/ Alamy Stock Photo.*

CHAPTER 4: BODIES SITUYATED IN NATURAL WORLDS

Page 134: Mohammed Shah's tomb, Lodhi Gardens, Delhi, India. © *Tibor Bognar/AGE Fotostock/VIEW.*

Page 139: Residents of housing projects thrive when nature is visible and accessible, and don't when it isn't: Ida B. Wells Housing, Chicago, Illinois [demolished]. *Chicago Architectural Photographing Company, ca. 1950s. CP-C_01_C_0265_004, University of Illinois at Chicago Library, Special Collections.*

Page 141: Affordable, sustainable, green, humane: Via Verde Housing (Grimshaw and Dattner), Bronx, New York. © *David Sundberg/Esto.*

Page 143: In healing gardens, heart rates settle and cortisol levels fall within minutes: Crown Sky Garden, Lurie Children's Hospital (Mikyoung Kim Design), Chicago, Illinois. *Photography by Hedrich Blessing/courtesy of Mikyoung Kim Design.*

Page 144: Workplaces with ample natural light boost job satisfaction: conference room, Scottish Parliament (Enric Miralles/EMBT Architects), Edinburgh, Scotland. © *Peter Cook/VIEW.*

Page 146: Views and paths invite us to explore unknown places: Connecticut Water Treatment Facility (Michael Van Valkenburgh Associates, building by Steven Holl). *Photograph by Elizabeth Felicella.*

Page 149: What's behind that concrete monolith of a wall? South facade, Salk Institute for Biological Studies (Louis Kahn), La Jolla, California. *Louis I. Kahn Collection, University of Pennsylvania and the Pennsylvania Hisorical and Museum Commission.*

Page 150, top: Kahn designed the Salk Institute with "a deep reverence for the nature of nature": view from Pacific Ocean, Salk Institute. © *Ezra Stoller/Esto.*

Page 150, bottom: Original east entrance through screen of eucalyptus trees, Salk Institute. © *Ezra Stoller/Esto.*

Page 152: Directing our eyes to the horizon and minimizing distractions: Central plaza with fountain, Salk Institute. *John Nicolais Collection, The Architectural Archives, University of Pennsylvania. Photo by John Nicolais, March 1979.*

Page 155: Some geons and their variants.

Page 156, top: Identifying geons: prisms and pyramids (compare with Heydar Aliyev): Ypenburg Housing (MVRDV), the Hague, Netherlands. © *Brian Rose.*

Page 156, bottom: Searching for geons (compare with Ypenburg Housing): Heydar Aliyev (Zaha Hadid), Baku. *Photo: Iwan Baan.*

Page 162: Perception is perception for action: two means of ascent at the Villa Savoye (Le Corbusier), Poissy, France. *akg-images/L. M. Peter.*

Page 164, top: Teasing gravity: unbuilt project for the Museum of Modern Art (Oscar Niemeyer), Caracas, Venezuela.

Page 164, bottom: Taunting gravity: cantilever at CCTV Headquarters (Rem Koolhaas/Office of Metropolitan Architecture), Beijing, China. *Nikolas Koenig/ OTTO.*

Page 167: A sudden attentional shift, from nature to culture: standing in the Salk Institute's central plaza, staircase-office blocks. *Photo: Iwan Baan.*

Page 168: "A building is a struggle, not a miracle. The architecture should acknowledge this": detail of concrete V joints, Salk Institute. © *Ezra Stoller/ Esto.*

Page 170, top: National Pensions Institute (Alvar Aalto), Helsinki, Finland: entrance is to left of the front block: *Photo by Heikki Havas, Alvar Aalto Museum. Circa 1957.*

Page 170, bottom: Suburban office building, United States. © *Anton Grassl/Esto.*

Page 171: Disaggregated volumes mitigate scale, National Pensions Institute. *Photo by Hekki Havas, Alvar Aalto Museum 1957.*

Page 172: Red rock, red brick: National Pensions Institute (detail, near entrance). *Sarah Williams Goldhagen.*

Page 173: Courtyard with garden, National Pensions Institute. *Photo: Maija Holma, Alvar Aalto Museum. 1997.*

Page 174, left: Double-tier skylights, National Pensions Institute. *Photo: Richard Peters, Alvar Aalto Museum. 1970s.*

Page 174, right: Baton tiles (interior) National Pensions Institute. *Photo: Maija Holma, Alvar Aalto Museum. 1997.*

Page 176: Accommodating climates and building cultures: Primary School (Diébédo Francis Kéré), Gando, Burkina Faso. © *Grant Smith/VIEW.*

Page 178: High-rise nature: Vertical Forest (Stefano Boeri), Milan, Italy. *Photography* © *Davide Piras/Courtesy of Boeri Studio.*

Page 179: Tall buildings in warm climates: Newton Suites (WOHA Architects), Singapore. *Photography by Patrick Bingham-Hall.*

Page 180: Qinhuangdao Habitat (Moshe Safdie), China. *Photography by Tim Franco and courtesy Safdie Architects.*

CHAPTER 5: PEOPLE EMBEDDED IN SOCIAL WORLDS

Page 184: Below: Panthéon (Jacques-Germain Soufflot), Paris, France. © *Honzahruby/Dreamstime.com.*

Page 185: Opposite: Interior, Panthéon. © *Dennis Dolkens/Dreamstime.com.*

Page 187, top: Historic photo of Jaffa Gate, Old City, Jerusalem. *Library of Congress, Prints and Photographs Division_ LC-DIG-ppmsca-02688.*

Page 187, bottom: A riot of visual and auditory stimuli: souk, Old City, Jerusalem. *Neil Farrin/AWL Images/Getty Images.*

Page 190: Insa-dong District, Seoul, South Korea. *Sarah Williams Goldhagen.*

Page 191: Ssamziegil (Moongyu Choi + Ga.A), Insa-dong District, Seoul. *Sarah Williams Goldhagen.*

Page 192: Opposite: Entrance, Ssamziegil. *Sarah Williams Goldhagen.*

Page 193: Above: Shops edge a gently inclined ramp, Ssamziegil. *Sarah Williams Goldhagen.*

Page 193: Below: Textures, patterns, and materials in hanok facades, Bukchon District, Seoul. © *Paul Brown/AGE Fotostock/VIEW.*

Page 195: Teahouse in a renovated hanok, Bukchon. *Sarah Williams Goldhagen.*

Page 201: Home schema: Tokyo Apartments (Sou Fujimoto), Tokyo, Japan. © *Edmund Sumner/VIEW.*

Page 203: Home schema: Vitra showroom (Herzog & de Meuron), Weil-am-Rhein, Germany. *akg-images/VIEW Pictures.*

Page 206: Action settings offer multiple uses: National Opera and Ballet (Snøhetta), Oslo, Norway. *Ferry Vermeer/Moment/Getty Images.*

Page 208: Gund Hall (John Andrews), main facade, Harvard University, Cambridge, Massachusetts. *Photo courtesy Harvard University Graduate School of Design.*

Page 209: Gund Hall, rear facade (trays housed in pitched skylit shed at left). *Photo courtesy Harvard University Graduate School of Design.*

Page 213: Foyer, Gund Hall. *Photo courtesy Harvard University Graduate School of Design.*

Page 214: The trays, Gund Hall. *Photo courtesy Harvard University Graduate School of Design.*

Page 215: Few places to gather: professor instructing students in Gund Hall's trays. *Photo courtesy Harvard University Graduate School of Design.*

Page 221: Without patterns, chaos: Rehak House (Coop Himmelb(l)au), unbuilt. © *Tom Bonner.*

Page 223: Geons in buildings. © *Edmund Sumner/VIEW.*

Page 224,: Fractals in buildings: Kandariya Mahadeva Temple, Khajuraho, India. *Yvan Travert/akg-images.*

Page 225: Gateway (Propylaea) to the Acropolis, with the Parthenon in the distance, Athens, Greece. *akg-images/De Agostini/Archivio J. Lange.*

Page 226: Reconstruction of the site plan, Acropolis: Propylaea is columned structure at left, Parthenon is bottom right. *The Granger Collection, New York.*

Page 227: Caryatids function as columns: Erectheion, Acropolis. *haris vithoulkas/Alamy Stock Photo.*

Page 228: Approaching the Parthenon, Acropolis. *akg-images/De Agostini/A. Vergani*

Page 230: Mansudae Assembly Hall, Pyongyang, North Korea. *Roman Harak.*

Page 232: Illustration exaggerating the optical refinements in the Parthenon. *Wikipedia.*

Page 234: Below: The Federal Treasury, Washington, DC. *Rob Crandall/Alamy Stock Photo*

Page 235: Opposite: Temple of Athena Nike, Acropolis. *DEA/G. DAGLI ORTI/ De Agostini/Getty Images*

Page 237: Chunking by function or structure: Schnabel House (Frank Gehry), Brentwood, California. © *2016 Nick Springett Photography.*

Page 238, top: Francis A. Gregory Neighborhood Library (David Adjaye), Washington, DC. © *Edmund Sumner/VIEW.*

Page 238, bottom: Patterned complexity in the interior of Francis A. Gregory Library. © *Edmund Sumner/VIEW*.

Page 240: Opposite: GSW Headquarters (Sauerbruch Hutton), Berlin, Germany. *Arco/Schoening/AGE Fotostock*.

Page 241: Below: St. Nicholas Orthodox Christian Church (Marlon Blackwell), Springdale, Arkansas. *Tim Hursley*.

Page 243, top: Foyer, St. Nicholas Church. *Tim Hursley*.

Page 243, bottom: Sanctuary, St. Nicholas Church. *Tim Hursley*.

Page 244: Below: The interaction of materials in light: Andy Goldsworthy, *Sticks Laid This Way, Then That* (detail), photographed at three different times in one day. *Andy Goldsworthy Courtesy Galerie Lelong, New York*.

Page 245: Opposite: Materials and texture in light: Exterior wall detail, Clyfford Still Museum (Brad Cloepfil), Denver, Colorado. *Raul J Garcia/arcaidimages.com*.

Page 246: Materials, light, geometry, texture: Interior, Clyfford Still Museum. *Pygmalion Karatzas/arcaidimages.com*.

Page 248: Changes in climate activate the site: view from the ocean, the Sea Ranch, California. *Eros Hoagland/Redux*.

Page 249: Typical house, the Sea Ranch. *Walter Bibikow/AGE footstock/VIEW*.

Page 250: Leça Swimming Pools (Alvaro Siza), Porto, Portugal. *Serrat/Alamy Stock Photo*.

Page 251: Wright envisioned the Guggenheim Museum as a place where people would gather to look at art and each other—while the kids entertained themselves. *The Frank Lloyd Wright Foundation Archives (The Museum of Modern Art / Avery Architectural & Fine Arts Library, Columbia University, New York) /© 2016 Frank Lloyd Wright Foundation, Scottsdale, AZ/Artists Rights Society (ARS), NY*.

Page 252: The Palais Garnier foyer, with multiple staircases and landings, was designed as a place to see others and be seen (Charles Garnier), Paris, France. *Library of Congress, Prints and Photographs Division, LC-DIG-ppmsc-09977*.

Page 253: Villa dall'Ava (Rem Koolhaas/Office of Metropolitan Architecture), Paris, France. *Photograph by Hans Werlemann, Courtesy of OMA*.

Page 254: The benches at the High Line turn people into part of the composition (James Corner Field Operations with Diller Scofidio + Renfro), New York, New York. *Photo: Iwan Baan*.

Page 256: Metropol Parasol: trees (or giant mushrooms) shading a large plaza in central Seville (Jurgen Mayer H.), Spain. *Julian Castle/arcaidimages.com*.

Page 257: Sails, shells, and soaring: metaphors at the Sydney Opera House (Jørn Utzon), Sydney, Australia. *Michael Weber/imageBROKER/AGE Fotostock*.

Page 258: Bird's Nest Stadium (Herzog & de Meuron), Beijing, China. © *Shu He/VIEW*.

Page 261: Scottish Parliament (Enric Miralles/EMBT Architects), Edinburgh, Scotland. © *Roland Halbe.*

Page 262: Petal chamber, Scottish Parliament. © *Peter Cook/VIEW.*

Page 263: Exterior office blocks with projecting reading nooks, Scottish Parliament. © *Peter Cook/VIEW.*

Page 264: Reading nook in each Member of Parliament's office. © *Roland Halbe.*

Page 266: Assembly Chamber, Scottish Parliament. © *Keith Hunter/arcaidimages.com.*

CHAPTER 7: FROM BLINDSIGHT TO INSIGHT

Page 273: Inequality is inscribed into our landscapes: modern apartments adjacent to slum dwellings, Mumbai, India. *Dinodia Photos/Alamy Stock Photo.*

Page 275: "Non-place" places: cityscape, China. *Photo by Kai M. Caemmerer.*

Page 276: Some housing shouts in the face of occupants that their lives don't matter: East Harlem, New York. *Sarah Williams Goldhagen.*

Page 279: Positive freedoms: "half a house" to start: Quinta Monroy Housing (Alajandro Aravena/ELEMENTAL), Iquique, Chile. *photo: Cristobal Palma/Estudio Palma.*

Page 280: Modular sustainable housing (Tatiana Bilbao Studio), Chiapas, Mexico. *Tatiana Bilbao Estudio.*

Page 283: Enriched environments: 798 Art District, Beijing, China. *Sarah Williams Goldhagen.*

Page 289: Aqua Building (Studio Gang), Chicago, Illinois. *WilsonsTravels Stock/Alamy Stock Photo.*

Page 291, top: School of Design (NADAAA with John Wardle), University of Melbourne, Melbourne, Australia. *Photography by John Horner/Melbourne School of Design, John Wardle Architects and NADAAA in collaboration.*

Page 291, bottom: Playing at Geopark: Geopark (Helen & Hard), Stavanger, Norway. *Photography by Emile Ashley/Courtesy Helen & Hard.*

Notes

xvi **The built environment conveys** Data in this paragraph are drawn and calculated from the United States Census Bureau, Projections of the Size and Composition of the U.S. Population: 2014 to 2060 Population Estimates and Projections Current Population Reports, by Sandra L. Colby and Jennifer M. Ortman (March 2015); census.gov/popclock/; en.wikipedia .org/wiki/List_of_metropolitan_areas_of_the_United_States; Jennifer Seal Cramer and William Browning, "Transforming Building Practices trhough Biophilic Design," ed. Stephen R. Kellert, Judith Heerwagen, and Martin L. Mador, ed., *Biophilic Design: The Theory, Science, and Practice of Bringing Buildings to Life* (New York: Wiley, 2008), 335–46.

xviii **By 2030,** Arthur C. Nelson, "Toward a New Metropolis: The Opportunity to Rebuild America," Brookings Institution (2004), brookings.edu/~/media/ research/files/reports/2004/12/metropolitanpolicy-nelson/20041213_ rebuildamerica.pdf.

xviii **Around the globe, a little more than** United Nations un.org/esa/popula tion/publications/sixbillion/sixbilpart1.pdf. One study helps make sense of the global trends toward hyper urbanization: Shlomo Angel, *Planet of Cities* (Cambridge, MA: Lincoln Institute of Land Policy, 2012).

xviii **And the number of megalopolises** United Nations, Department of Economic and Social Affairs, Population Division, *World Urbanization Prospects: The 2014 Revision,* esa.un.org/unpd/wup/CD-ROM/.

xx **To accommodate the vast migration** "Preparing for China's Urban Billion," Global McKinsey Institute (March 2009); also see esa .un.org/wup2009/unup/index.asp?panel=1 (thanks to Christopher Rogacz for research assistance). Another way to calculate this would be to say that China needs to build one new city the size of the New York urban area every two years until 2030 http://special.globaltimes.cn.2010-11/ 597548 .html.

xxi **If you consider** Stephen Kellert, *Building for Life: Designing and Understanding the Human-Nature Connection* (Washington, DC: Island Press, 2005), 90–122; see also William A. Shutkin, *The Land That Could Be: Environmentalism and Democracy in the 21st Century* (Cambridge, MA: MIT Press, 2000).

xxvi **Churchill declared** "Churchill and the Commons Chamber," parliament .uk/about/living-heritage/building/palace/architecture.

xxvii **Certainly, some writers have led** The books cited in this paragraph, among the most influential books on urbanism of the last half century, are: Jane Jacobs, *The Death and Life of Great American Cities* (New York: Random House, 1961); Oscar Newman, *Defensible Space: Crime Prevention Through Urban Design* (New York: Macmillan, 1973); William H. Whyte, *The Social Life of Small Urban Spaces* (Ann Arbor, MI: Edwards Brothers, 1980); Jan Gehl, *Cities for People* (Washington, DC: Island Press, 2010) and *Life Between Buildings: Using Public Space* (Washington, DC: Island Press, 2011).

xxvii **Jacobs, Whyte, Newman and Gehl's:** Gaston Bachelard, *The Poetics of Space* (Boston: Beacon Press, 1964); Edward Casey, *Getting Back into Place: Toward a Renewed Understanding of the Place-World*, 2nd ed. (Bloomington, IN: Indiana University Press, 2009).

xxviii **Recently a group of cognitive neuroscientists** O'Keefe found the type of nerve cell in the brains of rats that also exists in the human brain that recognize place, while May-Britt and Edvard Moser identified the grid cells that facilitate wayfinding. See Marianne Fyhn et al., "Spatial Representation in the Entorhinal Cortex," *Science* 305, no. 5688 (August 27, 2004): 1258–264; Edvard I. Moser et al., "Grid Cells and Cortical Representation," *Nature Reviews Neuroscience* 15 (2014): 466–81; Edvard I. Moser et al., "Place Cells, Grid Cells, and the Brain's Spatial Representation System," *Annual Review of Neuroscience* 31 (2008): 69–89.

xxviii **Lynch's work highlighted** John B. Eberhard's *Brain Landscape: The Coexistence of Neuroscience and Architecture* (New York: Oxford University Press, 2008) and *Architecture and the Brain: A New Knowledge Base from Neuroscience* (Atlanta: Greenway, 2007) were among ANFA members' initial forays into the field (Eberhard is the organization's founder). Others: Harry Mallgrave, *The Architect's Brain: Neuroscience, Creativity, and Architecture* (New York: Wiley-Blackwell, 2010); Harry Mallgrave, *Architecture and Embodiment: The Implications of the New Sciences and Humanities for Design* (New York: Routledge, 2013); Juhani Pallasmaa, *The Eyes of the Skin* (Hoboken, NJ: Wiley, 2008); Sarah Robinson and Pallasmaa, eds., *Mind in Architecture: Neuroscience, Embodiment, and the Future of Design* (Cambridge, MA: MIT Press, 2013). See also Ann Sussman and Justin B. Hollander, *Cognitive Architecture: Designing for How We Respond to the Built Environment* (New York: Routledge, 2015). Two excellent studies that deal more extensively with urban design are Charles Montgomery, *Happy City:*

Transforming Our Lives Through Urban Design (New York: Farrar, Straus, 2013) and Colin Ellard, *Places of the Heart: The Psychogeography of Everyday Life* (New York: Bellevue, 2015).

xxix **An experience differs** John Dewey, *Art as Experience* (New York: Perigee Books, 1934): 37; Mark Johnson has an excellent discussion of Dewey, experience, and embodiment in "The Embodied Meaning of Architecture," ed. Robinson and Pallasmaa, *Mind in Architecture*, 33–50.

xxxi **Even in fully formed adults** Eleanor A. Maguire et al.; "Navigation-Related Structural Change in the Hippocampi of Taxi Drivers," *Proceedings of the National Academy of Sciences of the United States of America* 97, no. 8 (2000): 4398–403; Eleanor A. Maguire et al., "Navigation Expertise and the Human Hippocampus: A Structural Brain Imaging Analysis," *Hippocampus 13* (2003): 208–17.

xxxii **Once he put it this way** Kahn quoted in "Marin City Redevelopment," *Progressive Architecture* 41 (November 1960): 153; Joan Meyers-Levy and Rui Zhu, "The Influence of Ceiling Height: The Effect of Priming on the Type of Processing People Use," *Journal of Consumer Research* 34 (August 2007): 174–86.

CHAPTER 1: THE SORRY PLACES WE LIVE

1 **The headline to** Goldhagen, "Boring Buildings: Why Is American Architecture So Bad?" *American Prospect* (December 2001).

4 **Thirty percent of** UN Habitat, "Slums of the World: The Face of Urban Poverty in the New Millennium?" (2003) and "UN Habitat, The Challenge of Slums: Global Report on Human Settlements" (2003); Daniel Tovrov, "Five Biggest Slums in the World," *International Business Times* (December 2011), ibtimes.com/5-biggest-slums-world-381338; "5 Largest Slums in the World", http://borgenproject.org/5-largest-slums-world/.

4 **How might a child** Andrew Baum et al., "Stress and the Environment," *Journal of Social Issues* (January 1981): 4–35; Andrew Baum and G. E. Davis, "Reducing the Stress of High-Density Living: An Architectural Intervention," *Journal of Personality and Social Psychology* 38, no. 3 (1980): 471–81; Robert Gifford, *Environmental Psychology: Principles and Practices*, 5th ed. (Optimal Books, 2013), 253–54; Upali Nanda et al., "Lessons from Neuroscience: Form Follows Function, Emotion Follows Form," *Intelligent Buildings International* 5, suppl. 1 (2013): 61–78; Esther M. Sternberg and Matthew A. Wilson, "Neuroscience and Architecture: Seeking Common Ground," *Cell* 127, no. 2 (2006): 239–42.

10 **The difference in learning** Peter Barrett, Yufan Zhang, et al., "A Holistic, Multi-Level Analysis Identifying the Impact of Classroom Design on Pupils' Learning," *Building and Environment* 59 (2013): 678–79.

10 **"soft" classrooms** Gifford, *Environmental Psychology*, 330; C. Kenneth Tanner, "Effects of School Design on Student Outcomes," *Journal of*

Educational Administration 47, no. 3 (2009): 381–399; Rotraut Walden, ed., *Schools for the Future: Design Proposals from Architectural Psychology* (New York: Springer, 2015), 1–10: Walden writes that classrooms that worked best "did not look like learning spaces but rather like individualized, comfortable living rooms."

10 **Windowless rooms** John Zeisel, *Inquiry by Design: Environment/Behavior/Neuroscience in Architecture, Interiors, Landscape and Planning*, rev. ed. (New York: Norton, 2006), 12; Lisa Heschong, "An Investigation into the Relationship between Daylighting and Human Performance: Detailed Report," Heschong Mahone Group for Pacific Gas & Electric Company (1999); Simone Borrelbach, "The Historical Development of School Buildings in Germany," in ed. Walden, *Schools for the Future*, 51–88.

10 **And the sort of noise** Gifford, *Environmental Psychology*, 308–12.

12 **Explaining that he wanted** Nouvel, quoted in theguardian.com/artand design/2010/ jul/06/jean-nouvel-sepentine-pavilion.

12 **Humans respond to** Hessan Ghamari et al.; "Curved Versus Sharp: An MRI-Based Examination of Neural Reactions to Contours in the Built Healthcare Environment," conference paper, 2014; Oshin Vartanian et al., "Impact of Contour on Aesthetic Judgments and Approach-Avoidance Decisions in Architecture," *Proceedings of the National Academy of Sciences 110, suppl. 2* (2011): 10446–453. See studies also by Christian Rittelmeyer, cited in Rotraut Walden, ed., *Schools for the Future* (Springer, 2015), 98–99; Nancy F. Aiken, *The Biological Sources of Art* (Westport, CT: Praeger, 1998), 17.

12 **The color red** For an interesting discussion about how *seeing* red becomes the *experience of* seeing red, see Nicholas Humphrey, *Seeing Red: A Study in Consciousness* (Cambridge, MA: Harvard University Press, 2009).

12 **We know that** Sally Augustin, *Place Advantage: Applied Psychology for Interior Architecture* (Hoboken, NJ: Wiley, 2009), 142; Joy Monice Malnar and Frank Vodvarka, *Sensory Design* (Minneapolis: University of Minnesota, 2004), 205–6.

12 **An all-red environment** Esther M. Sternberg, *Healing Spaces: The Science of Place and Well-Being* (Cambridge, MA: Harvard University, 2009), 35–42.

17 **Our "report card"** infrastructurereportcard.org/a/#p/grade-sheet/gpa; nytimes.com/ 2014/01/15/business/international/indiasinfrastructure-projects-stalled-by-red-tape. html?_r=0.

18 **In much of Africa** Marianne Fay and Mary Morrison, "Infrastructure in Latin America and the Caribbean: Recent Developments and Key Challenges," The World Bank (Report number 32640, 2005).

19 **Contact with nature confers** Roger Ulrich, "Biophilic Theory and Research for Healthcare Design," in ed. Stephen R. Kellert, Judith Heerwagen, and Martin Mador, *Biophilic Design* (Hoboken, NJ: Wiley, 2008), 87–106; Kellert, "Dimensions, Elements, and Attributes of Biophilic Design," in *Biophilic Design*, 3–19; Sandra A. Sherman et al., "Post Occupancy Evaluation of Healing Gardens in a Pediatric Center" in ed. Cor Wagenaar, *The Architecture of Hospitals* (Rotterdam: Nai Publishers, 2006), 330–51.

19 **When people in** Commission for Architecture and the Built Environment (UK), *People and Places: Public Attitudes Toward Beauty* (2010).

19 **Merely the view of grass** Rachel Kaplan, "The Nature of the View from Home: Psychological Benefits," *Environment and Behavior* (2001): 507–42; Rachel Kaplan, "Environmental Appraisal, Human Needs, and a Sustainable Future," in ed. Tommy Gärling and Reginald G. Golledge, *Behavior and Environment: Psychological and Geographical Approaches* (Amsterdam: Elsevier, 1993): 117–40.

19 **And still, in spite** worldcitiescultureforum.com/data/of-public-green-space-parks-and-gardens.

21 **ambient noise levels** The Environmental Protection Agency set 55 dB as the standard for outdoor residential noise in the 1970s. The World Health Organization, in 1999, recommended 50 dB; World Health Organization, *Guidelines for Community Noise*, ed. Birgitta Berglund, Thomas Lindvall, and Dietrich H. Schwela, 1999. The information on New York City subways is from Lisa Goines and Louis Hagler, "Noise Pollution: A Modern Plague," *Southern Medical Journal* 100, no. 3 (2007): 287–94. In many cities, daytime noise levels hover around 75 decibels, even though levels above 60 dB are associated with higher rates of hospital admission for stroke: Jaana I. Halonen et al., "Long-term Exposure to Traffic Pollution and Hospital Admissions in London," *Environmental Pollution* 208, part A (2016): 48–57. In London, and presumably in other similarly large, world-class cities, decibel levels of 50–55 can be found only inside urban parks and interior courtyards, or in the dead of night; info.acoustiblok.com/blog/bid/70023/Noise-Pollution-Ranking-America-s-Noisiest-Cities; the atlanticcities.com/neighborhoods/2012/05/just-how-bad-noise-pollution-our-health/2008/.

22 **The European Union** Malnar and Vodvarka, *Sensory Design*, 138–39; D. Balogh et al., "Noise in the ICU," *Intensive Care Medicine* 19, no. 6 (1993): 343–46;

22 **The World Health Organization outlines** Baum et al., "Stress and the Environment," 23–25; Malnar and Vodvarka, *Sensory Design*, 138–39.

22 **Noise levels higher** I. Busch-Vishniac et al., "Noise Levels in Johns Hopkins Hospital," *Journal of the Acoustical Society of America* 118, no. 6 (2005): 3629–645.

22 **Exposure to continuous** WHO, *Guidelines for Community Noise*, 47–49.

22 **Children who attend** Arline L. Bronzaft, "The Effect of a Noise Abatement Program on Reading Ability," *Journal of Environmental Psychology* 1, no. 3 (1981): 215–22; cited in Gifford, *Environmental Psychology*, 309.

24 **70 percent of the houses** Ellen Dunham-Jones and June Williamson, *Retrofitting Suburbia: Urban Design Solutions for Redesigning Suburbs* (Hoboken, NJ: Wiley, 2009), 17, 235.

24 **The logic driving** An excellent introduction to the financial orientation and design methods of the US home-building industry is Anthony Alofsin, *Dream Home: What You Need to Know Before You Buy* (CreateSpace

Independent Publishing Platform, 2013). And on the US construction industry's egregious irrationalities and market inefficiencies—which greatly contribute to the multiple disincentives to innovate as well as to the appalling craftsmanship of most new American construction—see Barry B. LePatner, *Broken Buildings, Busted Budgets: How to Fix America's Trillion-Dollar Construction Industry* (Chicago: University of Chicago, 2008).

24 **Outdated land-use ordinances** On the inefficiency and anachronism of US zoning codes, see Edward Glaeser, *Triumph of the City: How Our Greatest Invention Makes Us Richer, Smarter, Greener, Healthier, and Happier* (New York: Penguin, 2011). On the inefficiency and egregious anachronism of both municipal building and zoning codes, see Jonathan Barnett, "How Codes Shaped Development in the United States, and Why They Should Be Changed," in ed. Stephen Marshall, *Urban Coding and Planning* (New York: Routledge, 2011), 200–26.

25 **developments promote lifestyles** Richard J. Jackson with Stacy Sinclair, *Designing Healthy Communities* (Hoboken, NJ: Wiley, 2012), 10; see also Jackson's website: designinghealthycommunities.org/.

25 **An auto-bound, sedentary lifestyle** Michael Mehaffey and Richard J. Jackson, "The Grave Health Risks of Unwalkable Communities," *Atlantic Cities,* citylab.com/design/2012/06/grave-health-risks-unwalkable-communities/2362/.

25 **suburbs can cultivate** Robert D. Putnam, *Bowling Alone: The Collapse and Revival of American Community* (New York: Simon & Schuster, 2000); Charles Montgomery, *Happy City*, 146–75.

25 **They lose out** Montgomery, *Happy City*, 227–50.

27 **nature tamed by suburbia** On how understimulation in both landscapes and architecture causes stress detrimental to health and well-being, see Henk Staats, "Restorative Environments," in ed. Susan D. Clayton, *Oxford Handbook of Environmental and Conservation Psychology* (New York: Oxford University Press, 2012), 445–58; Colin Ellard, *Places of the Heart*, 107–8; V. S. Ramachandran, *The Tell-Tale Brain: A Neuroscientist's Quest for What Makes Us Human* (New York: Norton, 2011), 218–44; Montgomery, *Happy City*, 91–115; and most recently, a synthesis of the basic research: Jacoba Urist, "The Psychological Cost of Boring Buildings," *New York,* April 2016: nymag.com/scienceofus/2016/04/the-psychological-cost-of-boring-buildings.html.

27 **Theo, the protagonist** Donna Tartt, *The Goldfinch* (New York: Little, Brown, 2013), 221–22.

27 **the same sort of tedium** On the multisensory nature of how people perceive surface-based cues. One study on how people perceived wood found that subjects were quite adroit at distinguishing real wood from fake: Krista Overvliet and Salvador Soto-Faraco, "I Can't Believe This Isn't Wood! An Investigation in the Perception of Naturalness," *Acta Psychologica* 136, no. 1 (2011): 95–111.

29 **Big Dig** Erich Moskowitz, "True Cost of Big Dig Exceeds $24 Billion with Interest, Officials Determine," *Boston.com* (July 10, 2012), boston. com/metrodesk/2012/07/10/true-cost-big-dig-exceeds-billion-with-inter est-officials-determine/AtR5AakwfEyORFSeSpBn1K/story.html.

30 **Six years after** Casey Ross, "Greenway Funds Fall Short as Costs Rise," *Boston.com*, April 19, 2010; boston.com/business/articles/2010/04/19/ greenway_hit_by_rising_costs_drop_in_state_funds/. See also Sarah Williams Goldhagen, "Park Here," *New Republic*, October 6, 2010.

30 **What's more, construction** See LePatner, *Broken Buildings*; Steven Kieran and James Timberlake, *Refabricating Architecture: How Manufacturing Methodologies Are Poised to Transform Building Construction* (New York: McGraw-Hill, 2003), 23.

32 **Consider, for example** On One WTC, see Sarah Williams Goldhagen, *Architectural Record,* January 2015.

34 **architectural education** The best critique of contemporary architectural education is Peter Buchanan's *The Big Rethink: Rethinking Architectural Education,* published as a series in the London-based *Architectural Review* in 2011 and 2012, and available online. Some of Buchanan's suggestions for the reconfiguration of architectural education echo those advanced here. In addition, D. Kirk Hamilton and David H. Watkins, in *Evidence-Based Design for Multiple Building Types* (Hoboken, NJ: Wiley, 2008), lay out a sensible case for modifying the way architects approach design vis-à-vis their clients and the marketplace, 1–26. Evidence-based design (EBD) is mainly used in buildings with health-care uses (although Hamilton and Watkins argue that it should pertain to a wider range of institutions, such as schools and workplaces). The standards for EBD are and should be stringent. The approach proposed here can include EBD principles, but can and should extend beyond their rather limited province.

Christopher Alexander is perhaps one of the earliest and certainly the best-known (to architects, at least) proponents of an EBD approach: *A Pattern Language: Towns, Buildings, Construction* (New York: Oxford University Press, 1976), although what Alexander considers "evidence" would not meet contemporary standards. *A Pattern Language* and Alexander's subsequent books present a puzzling mix of trenchant observations, commonsense recommendations, and retrograde, antimodern tenets for design. And nowhere does Alexander deal systematically with irrationalities in human perception and cognition, which are legion.

34 **Unusual forms and statement architecture** Buchanan, "The Big Rethink I," *Architectural Review* (2012), complains that at too many major schools of architecture, "rather than relevance, what is sought out is originality, no matter how spurious"; see also David Halpern, "An Evidence-Based Approach to Building Happiness," in *Building Happiness: An Architecture to Make You Smile,* ed. Jane Wernick (London: Black Dog, 2008), 160–61.

34 **Jeff Speck** Speck quoted in Martin C. Pedersen, "Step by Step, Can American Cities Walk Their Way to Healthy Economic Development?," *Metropolis,* October 2012: 30. A sensitive assessment of how students are becoming less well trained in the essential tool of being able to accurately assess scale, see also Tim Culvahouse, "Learning How Big Things Are," at tculvahouse.tumblr.com/post/123316363707/learning-how-big-things-are.

36 **Photographs confer the impression** On some of the ways that photography distorts built environments in general and architectural space in particular, see these articles, all by Claire Zimmerman: "Photography into Building in Postwar Architecture: The Smithsons and James Stirling," *Art History,* April 2012: 270–87; "The Photographic Image from Chicago to Hunstanton," in ed. M. Crinson and C. Zimmerman, *Neo-avant-garde and Postmodern: Postwar Architecture in Britain and Beyond* (New Haven: Yale University Press, 2010), 203–28; "Photographic Modern Architecture: Inside 'The New Deep,'" *Journal of Architecture* 9, no. 3 (2004): 331–54; "The Monster Magnified: Architectural Photography as Visual Hyperbole," *Perspecta* 40 (2008): 132–43.

38 **photographs of another widely celebrated** Lawrence Cheek, "On Architecture: How the New Central Library Really Stacks Up," *Seattle Post-Intelligencer,* March 26, 2007.

39 **Most surprisingly** Antonio Damasio, in *The Feeling of What Happens: Body and Emotion in the Making of Consciousness* (New York: Harcourt Brace, 1999), 1–51, establishes (following William James) that human emotions are first and foremost body-based. He writes, "emotions use the body as their theater" (51), explaining that emotions are changes in our internal body milieu in response to either external experiences or our internal representations of those experiences. One of dozens of articles emphasizing the relationship of human emotions to spatial navigation and other aspects of environmental perception, see Elizabeth A. Phelps, "Human Emotions and Memory: Interactions of the Amygdala and Hippocampal Complex," *Current Opinion in Neurobiology* 14 (2004): 198–202.

39 **We also mostly ignore** Anthony Giddens discusses the reliance on expertise as one of the conditions and consequences of modern life in *The Consequences of Modernity* (Stanford, CA: Stanford University Press, 1991).

40 **Not only are consumers** Processing anomalous information takes much more energy than processing typical information: William R. Hendee and Peter N. T. Wells, *The Perception of Visual Information* (New York: Springer, 1997). Daniel Kahneman, in *Thinking, Fast and Slow* (New York: Farrar, Straus, 2010), 66, discusses the dynamics of anomaly detection in his description of the "mere exposure effect"; and describes how it disposes people to deem something as objectively normative, 103. Some research robustly confirms the importance of the mere exposure effect to the development of place attachment; others cast doubt upon its relevance:

in a meta-analysis of ten studies, the mere exposure effect's pertinence to place attachment was shown to be moderate or even weak: Kavi M. Korpela, "Place Attachment," in *Oxford Handbook of Environmental and Conservation Psychology*, 152.

40 **In the wake** Steven Pinker, for example, perpetuates this conventional view, maintaining that art constitutes nothing more than a device for "pushing our pleasure buttons": *How the Mind Works* (New York: Norton, 1997), 539; although to be sure, the comment is not about the built environment but artistic practice more generally. For a more enlightened approach, see V. S. Ramachandran and W. R. Hirstein, "The Science of Art: A Neurological Theory of Aesthetic Experience," *Journal of Consciousness Studies* 6, nos. 6–7 (1999), 15–51; and Ramachandran, *The Tell-Tale Brain*, 241–45.

CHAPTER 2: BLINDSIGHT

44 **For understanding how** Blindsight is a well-known phenomenon in cognitive neuroscience and studies of visual perception, especially because it poses salient issues for the understanding of consciousness. For one discussion, see Güven Güzeldere et al., "The Nature and Function of Consciousness: Lessons from Blindsight," *The New Cognitive Neurosciences*, 2nd ed., ed. Michael S. Gazzaniga (Cambridge, MA: MIT Press, 2000), 1277–284. The study of the subject with left hemisphere neglect and the burning house is recounted in John C. Marshall and Peter W. Halligan, "Blindsight and Insight in Visuo-Spatial Neglect," *Nature* 336, no. 6201 (1988): 766–67.

46 **If someone sitting** Angela K.-y. Leung et al., "Embodied Metaphors and Creative 'Acts,'" *Psychological Science 23* (2012): 502–9.

46 **Or this: if your** Michael L. Slepian et al.; "Shedding Light on Insight: Priming Bright Ideas," *Journal of Experimental Social Psychology* 46, no. 4 (2010): 696–700; for the impact of bright light on emotional affect, see Alison Jing Xu and Aparna A. Labroo, "Turning on the Hot Emotional System with Bright Light," *Journal of Consumer Psychology* 24, no. 2 (2014): 207–16.

46 **Or this: a real estate broker** Oshin Vartanian et al.; "Impact of Contour on Aesthetic Judgements and Approach-Avoidance Decisions in Architecture," *Proceedings of the National Academy of Science 10, suppl. 2* (2013): 10446–453: Ori Amir, Irving Biederman, and Kenneth J. Hayworth, "The Neural Basis for Shape Preference," *Vision Research* 51, no. 20 (2011): 2198–206.

47 **The new paradigm** Effectively I am introducing here the concept of embodied cognition, which is also sometimes called "grounded cognition" or, the term I prefer, "situated cognition." These integrally related but not completely overlapping concepts together help to constitute this new para-

digm of cognition, which increasingly is finding confirmation in cognitive neuroscience. The literature on embodied cognition is vast and growing. Some sources I have found particularly helpful are by Lawrence W. Barsalou and Mark Johnson. By Barsalou: "Grounded Cognition," *Annual Review of Psychology* 59 (2008): 617–45; "Grounded Cognition: Past, Present and Future," *Topics in Cognitive Science* 2 (2010): 716–24, and Barsalou et al., "Social Embodiment," in ed. Brian H. Ross, *The Psychology of Learning and Motivation: Advances in Research and Theory* 43 (2003): 43–92. Mark Johnson, *The Body in the Mind: The Bodily Basis of Meaning, Imagination, and Reason* (Chicago: University of Chicago, 1987); Johnson, *Meaning of the Body: Aesthetics of Human Understanding* (Chicago: University of Chicago, 2008), and Johnson with George Lakoff, *Philosophy in the Flesh: The Embodied Mind and Its Challenge to Western Thought* (New York: Basic Books, 1999).

Recent work in psychology and cognitive neuroscience increasingly confirms the embodied mind paradigm: examples can be found in *The Cambridge Handbook of Situated Cognition*, ed. Philip Robbins and Murat Aydede (New York: Cambridge University Press, 2009); *The Routledge Handbook of Embodied Cognition*, ed. Lawrence Shapiro (New York: Routledge, 2014); Raymond W. Gibbs Jr., *Embodiment and Cognitive Science* (New York: Cambridge University Press, 2005); Evan Thompson, *Mind in Life: Biology, Phenomenology, and the Sciences of the Mind* (Cambridge, MA: Harvard University Press, 2007).

48 **The emerging mind-body-environment paradigm** Barsalou, "Grounded Cognition," 619, 635, explains the relationship of grounded to embodied cognition—plato.stanford.edu/entries/embodied-cognition/#MetCog—and writes that still in 1998 there was "widespread skepticism about grounded cognition" but that now, it is much more generally accepted. Also Paula M. Niedenthal and Barsalou, "Embodiment in Attitudes, Social Perception, and Emotion," *Personality and Social Psychology Review* 9, no. 3: 184–211, at 186, write that "the main idea underlying all theories of embodied cognition is that cognitive representations and operations are fundamentally grounded in their physical context."

50 **Cognitive scientists of all stripes** Johnson, *Meaning of the Body*, 25–35.

51 **Nonverbal cognitions** W. Yeh and Barsalou, "The Situated Nature of Concepts," *American Journal of Psychology* 119, no. 3 (2006): 349–84.

52 **One of the cognitive revolution's** Antonio Damasio emphasizes the nonconscious and embodied nature of thought in many of his books, including *Descartes' Error: Emotion, Reason, and the Human Brain* (New York: G.P. Putnam's, 1994); *Self Comes to Mind: Constructing the Conscious Brain* (New York: Pantheon, 2010), and in the previously cited *The Feeling of What Happens*; see also George Engel, "The Need for a New Medical Model: A Challenge for Biomedicine," *Science* 196, no. 4286 (1977): 129–36.

52 **We remain, in Daniel Kahneman's words** Kahneman, *Thinking, Fast and Slow*, 24, 200.

54 **That's one kind** Mental simulation is extensively discussed in the literature on embodied cognition, as well as in the literature on mirror neurons: Barsalou, "Grounded Cognition," Barsalou, "Perceptual Symbol Systems," *Behavioral and Brain Sciences* 22 (1999): 577–660; Damasio, *Feeling of What Happens* and *Self Comes to Mind;* Anjan Chatterjee and Oshen Vartanian, "Neuroaesthetics," in *Trends in Cognitive Science* 18 (2014): 370–75; Vittorio Gallese and Corrado Sinigaglia, "What Is So Special about Embodied Simulation?" *Trends in Cognitive Science* 15, no. 11 (2011): 512–19, and Vittorio Gallese, "Being Like Me: Self-Other Identity, Mirror Neurons, and Empathy," in ed. Susan Hurley and Nick Chater, *Perspectives on Imitation: From Neuroscience to Social Science*, vol. 1 (Cambridge, MA: MIT Press, 2005), 108–18.

54 **That simulation, like most** Gabriel Kreiman, Christof Koch, and Itzhak Fried found that 88 percent of the neurons that fire selectively when an image is actually seen also fire when it is mentally imagined, or simulated: "Imagery Neurons in the Human Brain," *Nature* 408 (November 16, 2000): 357–361. Bruno Laeng and Unni Sulutvedt found that if people mentally simulate the experience of looking into a light, their pupils dilate, as though they were actually seeing that light: "The Eye Pupil Adjusts to Imaginary Light," *Psychological Science* 25, no. 1 (2014): 188–97. On multisensory and cross-modal perception, see also Mark L. Johnson, "Embodied Reason," in *Perspectives on Embodiment: The Intersections of Nature and Culture*, ed. Gail Weiss and Honi Fern Haber (New York: Routledge, 1999), 81–102. Sensorimotor cognition is described in Erik Myin and J. Kevin O'Regan, "Situated Perception and Sensation in Vision and Other Modalities: A Sensorimotor Approach," *Cambridge Handbook of Situated Cognition*, 185–97.

54 **Such schemas, innumerable** Barbara Tversky, for example, writes vis-à-vis human spatial perception that it is "people's enduring conceptions of the spatial world that they inhabit rather than the momentary internalized imagery of the current scene": "Structures of Mental Spaces: How People Think About Space," *Environment and Behavior* 35, no.1 (2003), 66–80. The neurological basis of simulation is discussed in Jean Decety and Julie Grèzes, "The Power of Simulation: Imagining One's Own and Other's Behavior," *Brain Research* 1079, no. 1 (2006): 4–14.

55 **Some of these are familiar** Johnson, *Meaning of the Body*, cited above.

55 **Haptic impressions** Harry Mallgrave, *The Architect's Brain*, 189–206.

58 **Nonconscious and conscious** Philip Merikle and Meredyth Daneman, "Conscious vs. Unconscious Perception," *The New Cognitive Neurosciences*, 2nd ed., ed. Michael S. Gazzaniga (Cambridge, MA: MIT Press, 2000), 1295–303. Daniel Kahneman, in *Thinking, Fast and Slow*, differentiates nonconscious from conscious thought by using the terminology "System

One" and "System Two," but he describes the mind moving from one to another through a kind of toggle-switch system, with System Two kicking in when System One fails to produce a cognition that comports with reality. As I explain in chapter 7, I prefer the concept of a spectrum, whereby nonconscious cognitions can become available to conscious cognitions depending upon a variety of circumstances. For the model of nonconscious and conscious cognition that I prefer, see Stanislas Dehaene, *The Cognitive Neuroscience of Consciousness* (Cambridge, MA: MIT Press, 2001).

59 **Like Glass's generative river** Barbara Tversky, in "Spatial Cognition," *Cambridge Handbook,* describes two ways of responding to environment: one is responding from perception; the other, responding from memory, 205.

59 **Even when we pay** Merikle and Daneman, "Conscious vs. Unconscious Perception," *The New Cognitive Sciences*; J. M. Ackerman, C. C. Nocera, and John A. Bargh, "Incidental Haptic Sensations Influence Social Judgments and Decisions," *Science* 328, no. 5986 (2010): 1712–715.

59 **The door, in other words** In E. S. Cross, A. F. Hamilton, and S. T. Grafton, "Building a Motor Simulation de Novo: Observation of Dance by Dancers," *NeuroImage* 31, no. 3 (2006): 1257–267, the authors had dancers watch the performance of an unfamiliar dance by other dancers, and identified the same neurons firing as those that would fire if they themselves were to dance: the dancers were mentally simulating the body movementss they would make were they themselves performing the dance.

60 **This in turn permeates** Lera Boroditsky and Michael Ramscar, "The Roles of Body and Mind in Abstract Thought," *Psychological Science* 13, no. 2 (2002): 185–89; Barbara Tversky, "Spatial Cognition," *Cambridge Handbook,* and Tversky, "The Structure of Experience," in ed. T. Shipley and J. M. Zachs, *Understanding Events* (Oxford: Oxford University Press, 2008), 436–64; Catherine L. Reed and Martha J. Farah, "The Psychological Reality of the Body Schema: A Test with Normal Participants," *Journal of Experimental Psychology: Human Perception and Performance* 21, no. 2 (1995): 334–43, and Catherine L. Reed, "Body Schemas," in A. Meltzoff and W. Prinz, eds., *The Imitative Mind* (Cambridge: Cambridge University Press, 2002), 233–43.

60 **We now know** Antonio Damasio, *The Feeling of What Happens,* 1–60.

61 **Today, psychological research** Pencil in mouth study: Paula M. Niedenthal, "Embodying Emotion," *Science* 316, no. 5827 (2007): 1002–5.

61 **In all, you assumed** Niedenthal, "Embodying Emotion"; Paula M. Niedenthal, Lawrence Barsalou et al., "Embodiment in Attitudes, Social Perception, and Emotion," *Personality and Social Psychology Review* 9, no. 3 (2005): 184–211.

62 **Another way to say** Barbara Tversky describes experiments by Sadalla and Staplan that confirm that when people travel on a route, turning cor-

ners increases their perception of the distance covered: Tversky, "Spatial Cognition," *Cambridge Handbook of Situated Cognition*, 207.

64 **contemporary skyscrapers** David Childs, personal communication with the author, December 2014.

64 **Ludwig Hilberseimer** Richard Pommer, *In the Shadow of Mies: Ludwig Hilberseimer: Architect, Educator, and Urban Planner* (New York: Rizzoli, 1988).

66 **Haussmann's replanning of Paris** Haussmann arranging the stars is quoted in T. J. Clark, *The Painting of Modern Life: Paris and the Art of Manet and His Followers* (New York: Knopf, 1984), 42. On the putatively dehumanizing aspects of the grid, see Alberto Pérez-Gomez's writings, beginning with *Architecture and the Crisis of Modern Science*.

66 **Today, studies on** T. Hafting et al., "Microstructure of a Spatial Map in the Entorhinal Cortex," *Nature 436* (2005): 801–6; Niall Burgess, "How Your Brain Tells You Where You Are," TED Talks, ted.com/talks/neil_burgess_how_your_brain_tells_you_where_you_are/transcript?language=en. All the neurological studies on spatial navigation discussed here, including those that identified place cells and grid cells, were conducted on laboratory rats, not humans. But it is commonly believed that the human spatial navigation system works in the same way.

67 **Wright eschewed the simple rectilinear** Neil Levine, "Frank Lloyd Wright's Diagonal Planning Revisited," in ed. Robert McCarter, *On and By Frank Lloyd Wright: A Primer of Architectural Principles* (New York: Phaidon, 2012), 232–63.

69 **mass-customized** Stephen Kieran and James Timberlake, *Refabricating Architecture: How Manufacturing Methodologies Are Poised to Transform Building Construction* (New York: McGraw-Hill, 2004).

69 **Direct responses** Nancy Aiken, *Biological Sources*, uses the more conventional behaviorist terminology, calling direct responses unconditioned and indirect ones unconditioned; Roger Ulrich, "Biophilia, Biophobia, and Natural Landscapes" in ed. Stephen Kellert and Edmund O. Wilson, *The Biophilia Hypothesis* (Washington, DC: Shearwater, 1993), 78–138.

69 **Anyone who has visited** Esther Sternberg, *Healing Spaces*, 51–74.

70 **These are compulsions** Sally Augustin, *Place Advantage*, 111–34.

70 **Understimulating environments** Ellard, *Places of the Heart*, 107–24; Jacoba Urist, "The Psychological Cost of Boring Buildings," *Science of Us* (April 2016).

70 **Some places elicit** Judith H. Heerwagen and Bert Gregory, "Biophilia and Sensory Aesthetics" in ed. Stephen R. Kellert, Judith H. Heerwagen, and Martin L. Mador, *Biophilic Design: The Theory, Science, and Practice of Bringing Buildings to Life* (Hoboken, NJ: Wiley, 2008), 227–41.

72 **Most famously, a certain hue** On color, see Brent Berlin and Paul Kay, *Basic Color Terms: Their Universality and Evolution* (Berkeley: University of California Press, 1970); Henry Sanoff and Rotraut Walden, "School Environments" in *Oxford Handbook of Environmental and Conservation*

Psychology, 276–94; Sternberg, *Healing Spaces,* 24–53; Augustin, *Place Advantage,* 48, 142; Adam Alter, *Drunk Tank Pink: And Other Unexpected Forces That Shape How We Think, Feel, and Behave* (New York: Penguin, 2013), 157–80.

72 **Maurice Merleau-Ponty** Maurice Merleau-Ponty, *Phenomenology of Perception,* trans. Colin Smith (New York: Routledge, 1962), 211.

73 **All these are instances** On metaphors and concrete experience, see Lawrence W. Barsalou, "Grounded Cognition," 617–45.

<?> **Metaphors are schemas** George Lakoff and Mark Johnson, *Metaphors We Live By* (Chicago: University of Chicago, 1980); Lera Boroditsky, "Metaphoric Structuring: Understanding Time through Spatial Metaphors," *Cognition* 75 (2000): 1–28; James Geary, *I Is an Other: The Secret Life of Metaphor and How It Shapes the Way We See the World* (New York: Harper, 2011); on metaphors in architecture, see my "Aalto's Embodied Rationalism," in ed. Stanford Anderson, Gail Fenske, and David Fixler, *Aalto and America* (New Haven: Yale University Press, 2012), 13–35, and Brook Muller, "Metaphor, Environmental Receptivity, and Architectural Design," unpublished.

76 **skewed fit** In *The Interpretation of Cultures,* Clifford Geertz writes that a metaphor generates "an incongruity of sense on one level"—in reality, no building is water-like—to produce "an influx of significance on the other"—swimming pools evoke playfulness, childhood, abandon, health, nature . . . (New York: Basic Books, 1973), 210. Thomas W. Schubert and Gün R. Semin, "Embodiment as a Unifying Perspective for Psychology," *European Journal of Social Psychology* 39, no. 7 (2009): 1135–141. On the aesthetic effect of exaggeration, see V. S. Ramachandran's notion of "peak-shift," presented in Ramachandran and Hirstein's "The Science of Art: A Neurological Theory of Aesthetic Experience."

76 **Like the "at home" example** "Important is big" and other similar metaphors are discussed in Lakoff and Johnson's *Metaphors We Live By* and in *Philosophy in the Flesh,* 47–87. On people's association of verticality with power: Thomas W. Schubert, "Your Highness: Vertical Positions as Perceptual Symbols of Power," *Journal of Personality and Social Psychology* 89, no. 1 (2005): 1–21; with divinity, Brian P. Meier et al., "What's 'Up' With God: Vertical Space as a Representation of the Divine," *Journal of Personality and Social Psychology* 93, no. 5 (2007): 699–710.

79 **The subjects holding** Joshua M. Ackerman, Christopher C. Nocera, and John A. Bargh, "Incidental Haptic Sensations Influence Social Judgments and Decisions," *Science* 328, no. 5986 (2010): 1712–715; Nils B. Jostmann, Daniël Lakens, and Thomas W. Schubert, "Weight as an Embodiment of Importance," *Psychological Science* 20, no. 9 (2009): 1169–174; Hans Ijzerman, Nikos Padiotis, and Sander L. Koole, "Replicability of Social-Cognitive Priming: The Case of Weight as an Embodiment of Importance," *SSRN Electronic Journal* (April 2013): n.p. Some psychologists

have had trouble replicating the results of the clipboard experiment, causing consternation in the field. My position is this: even if the findings of one or another experiment fails to be confirmed through replication, the existence of so many studies confirming the pervasiveness of metaphors in people's cognitive patterns is convincing.

83 **human memory** Eric R. Kandel, *In Search of Memory: The Emergence of a New Science of Mind* (New York: Norton, 2006).

83 **Recall a clear childhood memory** Kahneman, *Thinking Fast and Slow*, discusses the "mere context effect," which Gifford, *Environmental Psychology: Principles and Practice*, 307, calls the "familiar context effect."

84 **long-term memories that are autobiographical** Eric Kandel, *In Search of Memory*, 281–95; Barbara Maria Stafford, *Echo Objects: The Cognitive Work of Images* (Chicago: University of Chicago, 2007), 107–8. On the interrelationship of memory and emotions, see Antonio Damasio, *The Feeling of What Happens,* and Damasio, *Self Comes to Mind: Constructing the Conscious Brain* (New York: Pantheon, 2010), Elizabeth A. Phelps, "Human Emotion and Memory: Interactions of the Amygdala and Hippocampal Complex," *Current Opinion in Neurobiology* 14, no. 2 (2004): 198–202.

84 **place cells** Matthew A. Wilson, "The Neural Correlates of Place and Direction," in *The New Cognitive Neurosciences*, 2nd ed., ed. Michael S. Gazzaniga (Cambridge, MA: MIT Press, 2000), 589–600. Note that place cells contain both metric (allocentric) *and* contextual (allocentric and egocentric) information.

84 **Here, then, is another stunning fact** This is dramatically illustrated throughout Elena Ferrante's Neapolitan novels, in which the two main characters, Elena and Lila, frequently define and position their past, present, and future selves vis-à-vis the impoverished neighborhood in which they were raised.

CHAPTER 3: THE BODILY BASIS OF COGNITION

93 **embodied as actors** Embodied cognition has already been introduced in these notes and some of its basic sources cited; to them, add Linda B. Smith, "Cognition as a Dynamic System: Principles from Embodiment," *Developmental Review* 25 (2005): 278–98; Alan Costall and Ivan Leudar, "Situating Action I: Truth in the Situation," *Ecological Psychology* 8, no. 2 (1996): 101–10; Tim Ingold, "Situating Action VI: A Comment on the Distinction Between the Material and the Social," *Ecological Psychology* 8, no. 2 (1996): 183–87, and Tim Ingold, "Situating Action V: The History and Evolution of Bodily Skills," *Ecological Psychology* 8, no. 2 (1996): 171–82.

94 **More technically** Ramachandran, *Tell-Tale Brain*, 37, 86.

95 **body schemas** Catherine L. Reed, "What Is the Body Schema?," in ed. Andrew N. Meltzoff, *The Imitative Mind: Development, Evolution, and Brain Bases* (New York: Cambridge University Press, 2002), 233–43; Tversky, in

"Spatial Cognition," *Cambridge Handbook of Situated Cognition*, 201–16, also has a succinct account of the basic body schemas. Linda B. Smith's work on body schemas, cited above, emphasizes the development of body schemas through bodily motion.

96 **When architects calculate** Donald A. Norman, *The Design of Everyday Things*, rev. ed. and Norman, *Emotional Design: Why We Love (or Hate) Everyday Things* (New York: Basic Books, 2003), discussed how everyday objects can (and cannot) be designed to appeal to human nonconscious cognitions as well as to our allocentric bodies.

97 **A poignant example** Discussed in Richard Joseph Neutra, *Survival Through Design* (New York: Oxford University Press, 1954), 58.

98 **designed for people's *allocentric*** Alvar Aalto, "Rationalism and Man," in *Alvar Aalto in His Own Words*, ed. Alvar Aalto and Göran Schildt (New York: Rizzoli, 1998), 89–93.

100 **City planning theorists** Peter Calthorpe, *The Next American Metropolis: Ecology, Community, and the American Dream* (New York: Princeton Architectural Press, 1995), for example, proposes "pedestrian pockets" linked by public transit.

103 **"Interiors," he explains** Peter Zumthor, *Atmospheres* (Zurich: Birkhäuser, 2006), 29.

106 **astonishingly, "extremely rational"** Aalto, in ed. Aalto and Schildt, *In His Own Words*, 269–75.

106 **Installed in 2006** Dimensions from cityofchicago.org/city/en/depts/dca/ supp_info/millennium_park_-artarchitecture.html.

106 **Before entering this courtyard** Upali Nanda writes, "A person walking down the street sees practically nothing but the ground floor of buildings, the pavement, and what is going on in the street itself," in *Sensthetics: A Crossmodal Approach to Sensory Design* (Berlin: VDM Verlag Dr. Mueller, 2008), 57.

110 **One prominent neuroscientist** Marcello Constantini et al., "When Objects Are Close to Me: Affordances in the Peripersonal Space," *Psychonomic Bulletin and Review* 18, no. 2 (2011): 302–8; Alain Berthoz and Jean-Luc Petit, *The Physiology and Phenomenology of Action*, trans. Christopher Macann (Oxford: Oxford University Press, 2008), 49–57.

110 **It's almost as if** James J. Gibson, *The Senses Considered as Perceptual Systems*, rev. ed. (New York: Praeger, 1983); James J. Gibson, *The Ecological Approach to Visual Perception* (New York: Psychology Press, 1986); Berthoz and Petit, *Physiology and Phenomenology*, 2, 66. Anthony Chemero, "What We Perceive When We Perceive Affordances: A Commentary on Michaels," *Ecological Psychology* 13, no. 2 (2001): 111–16; Anthony Chemero, "An Outline of a Theory of Affordances," *Ecological Psychology* 15, no. 2 (2003): 181–95; Anthony Chemero, "Radical Empiricism through the Ages," review of Harry Heft, *Ecological Psychology in Context: James Gibson, Roger Barker, and the Legacy of William James's Radical Empiricism*,

Contemporary Psychology 48, no. 1 (2003): 18–21; Patrick R. Green, "The Relationship between Perception and Action: What Should Neuroscience Learn from Psychology?" *Ecological Psychology* 13, no. 2 (2001): 117–22, Keith S. Jones, "What Is an Affordance?," *Ecological Psychology* 15, no. 2 (2003): 107–14.

111 **This is true** "Thinking about the world is also an act and can be modified by attention," Berthoz and Petit, *Physiology and Phenomenology*, 51. See also Green, "The Relation between Perception and Action," *Ecological Psychology* 13, no. 2 117–122, and Boris Kotchoubey, "About Hens and Eggs: Perception and Action, Ecology and Neuroscience: A Reply to Michaels," *Ecological Psychology* 13, no. 2 (2001): 123–33.

111 **Not everything in** Marcello Constantini, "When Objects Are Close to Me," *Psychonomic Bulletin*, 302–8.

113 **Because of the embodied** The extent of near space, as we perceive it, scales with arm length: Matthew R. Longo and Stella F. Lourenco, "Space Perception and Body Morphology: Extent of Near Space Scales with Arm Length," *Experimental Brain Research* 177, no. 2 (2007): 285–90.

114 **But objects needn't** Fred A. Bernstein, "A House Not for Mere Mortals," *New York Times*, April 2008; nytimes.com/2008/04/03/garden/03destiny.html.

119 **"categories are containers"** Lakoff and Johnson, *Philosophy in the Flesh*, 51.

120 **Other embodied, intersensory** J. Decety and J. Grèzes, "The Power of Simulation: Imagining One's Own and Other's Behavior," *Brain Research* 1079, no. 1 (2006): 4–14; R. H. Desai et al., "The Neural Career of Sensory-Motor Metaphors," *Journal of Cognitive Neuroscience* 23, no. 9 (2011): 2376–86; Lakoff and Johnson, *Philosophy in the Flesh*, 20–21.

123 **tactile and visual cognition** Harry Mallgrave, *The Architect's Brain*, 189–206.

127 **humans are exquisitely sensitive** Daniel Levitin, *This Is Your Brain on Music: The Science of a Human Obsession* (New York: Dutton, 2006) contains much discussion of how sonic experience deeply influences our emotional states; R. Murray Shafer, *The Soundscape* (Merrimack, MA: Destiny Books, 1993).

127 **Cathedrals create highly unusual** Jean-François Augoyard and Henri Torgue, *Sonic Experience: A Guide to Everyday Sounds* (Montreal: Queen's-McGill University Press, 2006) in particular discuss the cutout effect. Barry Blesser and Linda-Ruth Salter, *Spaces Speak, Are You Listening?* (Cambridge, MA: MIT Press, 2009). See also Augoyard and Torgue, *Sonic Experience*; Mirko Zardini, ed., *Sense of the City: An Alternate Approach to Urbanism* (Montreal: Canadian Centre for Architecture and Lars Müller Publishers, 2005).

130 **But the Amiens interior's ubiquitous** Blesser and Salter, *Spaces Speak*, 89.

130 **Because of its vastness** The experience of awe also affects people's percep-

tion of time, seeming to slow it down: Melanie Rudd, Kathleen D. Vohs, and Jennifer Aaker, "Awe Expands People's Perception of Time, Alters Decision Making, and Enhances Well-Being," *Psychological Science* 23, no. 10 (2012): 1130–136. On how awe promotes prosocial thoughts and conduct, see Anna Mikulak, "All About Awe," *Association for Psychological Science Observer* (April 2015); psychologicalscience.org/index.php/publications/observer/2015/april-15/all-about-awe.html, and Paul K. Piff, "Awe, the Small Self, and Prosocial Behavior," *Journal of Personality and Social Psychology*, 108, no. 8 (2015): 883–99.

CHAPTER 4: BODIES SITUATED IN NATURAL WORLDS

133 **flooded with relief** Rachel and Stephen Kaplan first proposed the idea that people's evolutionary heritage made them biologically attuned to nature such that being in nature reduced stress by allowing us to replenish diminished attentional resources: Kaplan and Kaplan, *The Experience of Nature: A Psychological Perspective* (New York: Cambridge University Press, 1989), and Stephen Kaplan, "The Restorative Benefits of Nature: Toward an Integrative Framework," *Journal of Environmental Psychology* 15, no. 3 (1995): 169–82; Stephen Kaplan, "Aesthetics, Affect, and Cognition: Environmental Preference from an Evolutionary Perspective," *Environment and Behavior* 19, no. 1 (1987): 3–32. Dozens and dozens of subsequent studies have confirmed and refined Kaplan and Kaplan's Attention Restoration Theory (abbreviated as ART): see ed. Paul A. Bell et al., *Environmental Psychology*, 5th ed. (New York: Psychology Press, 2005).

137 **Another is that people** The "prospect and refuge" hypothesis was first advanced by Jay Appleton in *The Experience of Landscape* (Hoboken, NJ: Wiley, 1975). Although it relies on an outdated notion of human evolution (that humans evolved exclusively in the savannahs of East Africa), studies continue to confirm the powerful biologically based pull of prospect and refuge landscapes. See Judith H. Heerwagen and Gordon H. Orians, "Humans, Habitats, and Aesthetics," 138–72 in ed. Kellert and Wilson, *The Biophilia Hypothesis* (interestingly, the authors discuss gender variation in prospect and refuge preferences, with women preferring more refuge and men more prospect); see also Kellert, "Elements of Biophilic Design" and Ulrich, "Biophilia, Biophobia," in ed. Kellert, *Building for Life*, 129, 73–137. John Falk and John Balling, "Evolutionary Influence on Human Landscape Preference," in *Environment and Behavior* 42, no. 4 (2010): 479–93. For an accessible, updated presentation of current thinking on the varied landscapes that early humans inhabited, fostering our adaptability and immense cognitive flexibility, see Steven R. Quartz and Terrence J. Senjowski, *Liars, Lovers, and Heroes: What the New Brain Science Reveals about How We Become Who We Are* (New York: HarperCollins, 2002). For another discussion of prospect and refuge in architecture, see Grant Hildebrand, *The Wright Space: Pat-*

tern and Meaning in Frank Lloyd Wright's Houses (Seattle, WA: University of Washington Press, 1991).

138　**Even if systematic** Ulrich, "Biophilia, Biophobia," in ed. Kellert and Wilson, *The Biophilia Hypothesis,* 96; Colin Ellard, *Places of the Heart,* 29–51; Commission for Architecture and the Built Environment (UK), *People and Places.*

139　**Most astonishing, the children** Andrea Taylor et al., "Growing Up in the Inner City: Green Spaces as Places to Grow," *Environment and Behavior* 30, no. 1 (1998): 3–27. See also Rebekah Levine Coley, William C. Sullivan, and Frances E. Kuo, "Where Does Community Grow? The Social Context Created by Nature in Urban Public Housing," *Environment and Behavior* 29, no. 4 (1997): 488–94. Frances Kuo's website contains many other research studies on the effects of nature on cognition, emotion regulation and behavior, and so on: lhhl.illinois.edu/all.scientific.articles.htm.

139　**Dozens of subsequent studies** Michelle Kondo et al., "Effects of Greening and Community Reuse of Vacant Lots on Crime," *Urban Studies* (2015): 1–17; Austin Troy, J. Morgan Grove, and Jarlath O'Neil-Dunne, "The Relationship between Tree Canopy and Crime Rates across an Urban-Rural Gradient in the Greater Baltimore Region," *Landscape and Urban Planning* 106, no. 3 (2012): 262–70; Koley, Sullivan, and Kuo, "Where Does Community Grow? The Social Context Created by Nature in Urban Public Housing"; Frances E. Kuo and William C. Sullivan, "Environment and Crime in the Inner City: Does Vegetation Reduce Crime?," *Environment and Behavior* 33, no. 3 (2001): 343–67.

141　**Think about this** Commission for Architecture and the Built Environment (UK) *People and Places,* 24–42; Suzanne Nalbantian, *Memory in Literature: From Rousseau to Neuroscience* (New York: Palgrave Macmillan, 2003), 85–140.

142　**Within six months** Augustin, *Place Advantage,* 187–188; Rachel Kaplan, "The Role of Nature in the Context of the Workplace," *Landscape and Urban Planning* 26 (1993): 193–201; Rachel Kaplan, "The Nature of the View from Home: Psychological Benefits," *Environment and Behavior* 33, no. 4 (2001): 507–42; Ilknur Turkseven Dogrusoy and Mehmet Tureyen, "A Field Study on Determination of Preferences for Windows in Office Environments," *Building and Environment* 42, no. 10 (2007): 3660–668. Researchers found that doubling the normal rate of air ventilation in an office building was associated with a sharp spike in occupants' cognitive performance: Joseph G. Allen et al., "Associations of Cognitive Function Scores with Carbon Dioxide, Ventilation, and Volatile Organic Compound Exposures in Office Workers: A Controlled Exposure Study of Green and Conventional Office Environments," *Environmental Health Perspectives* (October 2015), online.

142　**Making workplace environments** Judith H. Heerwagen and Gordon H. Orians, "Adaptations to Windowlessness: A Study of the Use of Visual

Decor in Windowed and Windowless Offices," *Environment and Behavior* 18, no. 5 (1986): 623–39; Phil Leather et al., "Windows in the Workplace: Sunlight, View, and Occupational Stress," *Environment and Behavior* 30, no. 6 (1998): 739–62; Anjali Joseph, "The Impact of Light on Outcomes in Healthcare Settings," *Center for Health Design* issue paper #2, August 2006, healthdesign.org/chd/research/impact-light-outcomes-healthcare-settings; John Zeisel and Jacqueline Vischer, *Environment/Behavior/Neuroscience Pre & Post Occupancy of New Offices* (Society for Neuroscience, 2006).

142 **And these salutary physiological** Sandra A. Sherman et al., "Post Occupancy Evaluation of Healing Gardens in a Pediatric Center," in Cor Wagenaar, ed., *The Architecture of Hospitals,* 330–51: Several studies of well people (not patients) suggest even very short encounters with real or simulated natural settings trigger significant psychophysiological restoration, "within three to five minutes at most, or as quickly as twenty seconds."

145 **Retail environments that cater** Information in this paragraph from Heschong, "An Investigation," Heschong Mahone Group. See also Judith Heerwagen, "Investing in People: The Social Benefits of Sustainable Design," cce.ufl.edu/wpcontent/uploads/2012/08/Heerwagen.pdf; Phil Leather et al., "The Physical Workspace," in ed. Stavroula Leka and Jonathan Houdmont, *Occupational Health Psychology* (Hoboken, NJ: Wiley-Blackwell, 2010), 225–49; Zeisel and Vischer, *Environment/Behavior/Neuroscience*; Nanda and Pati, "Lessons from Neuroscience," ANFA presentation 2012.

146 **Children in properly daylit** Malnar and Vodvarka, *Sensory Design,* 199–228.

147 **Natural light, a boon** Jennifer A. Veitch, "Work Environments," in ed. Susan Clayton, *Oxford Handbook of Environmental and Conservation Psychology,* 248–75.

147 **These include ones** More recently, others have advanced a more complex scenario, relying on evidence of multiple changes in human habitats owing to unstable weather patterns and human migrations. See, for example, Quartz and Sejnowski, *Liars, Lovers, and Heroes.*

147 **Exposing people even** Upali Nanda, "Art and Mental Health," *Healthcare Design Magazine,* September 21, 2011.

151 **The visual field** Julian Hochberg, "Visual Perception in Architecture," *Via: Architecture and Visual Perception* 6 (1983): 27–45.

151 **As a result** Ellard, *Places of the Heart,* 37–46.

152 **It is as though** Kahn said, "One of the most wonderful buildings in the world which conveys its ideas is the Pantheon. The Pantheon really is a world within a world," in Louis I. Kahn, ed. Robert Twombley, *Louis I. Kahn: Essential Texts* (New York: Norton, 2003), 160; on the dependence of cognitions on emotions, see Damasio, *The Feeling of What Happens.*

153 **Kahn designed these approach** Kahn, quoted in H. F. S. Cooper, "The Architect Speaks," *Yale Daily News,* November 6, 1953, 2.

154 **Our mental representations** Semir Zeki, "The Neurology of Ambiguity," *Consciousness and Cognition 13* (2004): 173–96; Damasio, *Descartes' Error: Emotion, Reason, and the Human Brain* (New York: Penguin, 1994), 148–60; Harry Mallgrave, *Architecture and Embodiment,* 38–45.

154 **Because of the human eye's** Pinker, *How the Mind Works,* summarizing David Marr, 213.

154 **Scanning the Salk Institute** Distinction between form and surface-based cues in Tversky, "Spatial Thought, Social Thought," *Spatial Dimensions,* 20.

154 **Geons, in the words of the vision scientist** Irving Biederman, "Recognizing Depth-Rotated Objects: A Review of Recent Research and Theory," *Spatial Vision 13* (2001): 241–53; Biederman, "Recognition-by-Components: A Theory of Human Image Understanding," *Psychological Review* 94, no. 2 (1987): 115–47; O. Amir, Irving Biederman, and K. J. Hayworth, "The Neural Basis for Shape Preferences," *Vision Research 51,* no. 20 (2011): 2198–206.

155 **Geonic shapes abide** George Lakoff and Rafael Nuñez, *Where Mathematics Comes From: How the Embodied Mind Brings Mathematics into Being* (New York: Basic Books, 2000); Véronique Izard et al., "Flexible Intuitions of Euclidean Geometry in an Amazonian Indigene Group," *Proceedings of the National Academy of Sciences 108,* no. 24 (2011): 9782–787; Elizabeth Spelke, Sang Ah Lee, and Véronique Izard, "Beyond Core Knowledge: Natural Geometry," *Cognitive Science 34,* no. 5 (2010): 863–84; Berthoz and Petit, *Physiology and Phenomenology.* Giacomo Rizzolatti writes that no three-dimensional-shape geometrical is perceived as simply an abstract organization; instead, they "incarnate the practical opportunities that the object offers to the organism which perceives it," Anna Berti and Giacomo Rizzolatti, "Coding Near and Far Space," in ed. Hans-Otto Karnath, A. David Milner, and Giuseppe Valler, *The Cognitive and Neural Bases of Spatial Neglect* (New York: Oxford University Press, 2003), 119–29.

155 **Touching, even just seeing** Berthoz and Petit, *Physiology and Phenomenology,* 1–6; put differently, pure sensation "devoid of interpretation" does not exist, 48.

158 **This pathway suggests** The following analysis is based in part on the description of visual cognition advanced in Melvyn A. Goodale and David Milner, *Sight Unseen: An Exploration of Conscious and Unconscious Vision* (New York: Oxford University Press, 2004).

158 **Our responses to surfaces** On surface, see Jonathan S. Cant and Melvyn A. Goodale, "Attention to Form or Surface Properties Modulates Different Regions of Human Occipitotemporal Cortex," *Cerebral Cortex* 17, no. 3 (2007): 713–31.

158 **"Strange as it may seem"** Neutra, *Survival through Design,* 25.

159 **As Wright ushered** Malnar and Vodvarka, *Sensory Design*, 129–52.

160 **When a building's surfaces** Vittorio Gallese and Alessandro Gattara, "Embodied Simulation, Aesthetics, and Architecture," in ed. Sarah Robinson and Juhani Pallasmaa, *Mind in Architecture*, 161–79. The authors write on p. 164: "Embodied simulation can illuminate the aesthetic aspects of architecture . . . by revealing the intimate intersubjective nature of any creative act: where the physical object, the product of symbolic expression, becomes the mediator of an intersubjective relationship between creator and beholder."

160 **"Viewing hand-formed pottery"** Neutra, *Survival Through Design*, 74.

160 *Canonical neurons* **and** *mirror neurons* The information on mirror and canonical neurons in this and the following paragraphs: L. F. Aziz-Zadeh et al., "Lateralization in Motor Facilitation during Action Observation: A TMS Study," *Experimental Brain Research* 144, no. 1 (2002): 127–31; Damasio, *Self Comes to Mind*, 102–103; Erol Ahin and Selim T. Erdo An, "Towards Linking Affordances with Mirror/Canonical Neurons," unpublished (pdf); Vittorio Gallese and Alessandro Gattara, "Embodied Simulation, Aesthetics, and Architecture" (161–80) and Harry Francis Mallgrave, "Know Thyself: Or What Designers Can Learn from the Contemporary Biological Sciences" (9–31) in ed. Robinson and Pallasmaa, *Mind in Architecture*; David Freedberg and Vittorio Gallese, "Motion, Emotion and Empathy in Esthetic Experience," *Trends in Cognitive Science* 11, no. 5 (2007): 197–203; Giacomo Rizzolatti and Maddelena Fabbri Destro, "Mirror Neurons," *Scholarpedia* 3, no. 1 (2008): 2055. See also Eric Kandel, *The Age of Insight: The Quest to Understand the Unconscious in Art, Mind, and Brain, from Vienna 1900 to the Present* (New York: Random House, 2012), 418–20.

162 **For example, when a person** Lawrence E. Williams and John A. Bargh, "Experiencing Physical Warmth Promotes Interpersonal Warmth," *Science* 322, no. 5901 (2008): 606–7; Brian P. Meier et al., "Embodiment in Social Psychology," *Topics in Cognitive Science* (2012): 705–16. On the embodied metaphors underlying such associations, see Lakoff and Johnson, *Philosophy in the Flesh*, 45–46.

162 **A student will be** Joshua M. Ackerman, Christopher C. Nocera, and John A. Bargh, "Incidental Haptic Sensations Influence Social Judgments and Decisions," *Science* 328, no. 5986 (2010): 1712–715.

162 **Meet a new person** Siri Carpenter, "Body of Thought: Fleeting Sensations and Body Movements Hold Sway Over What We Feel and How We Think," *Scientific American Mind*, January 1, 2011: 38–45, 85.

163 **As humans grow** Pinker, *Mind*, 299–362.

165 **In our human** Johnson, *Meaning of the Body*, 160–61; Tversky, "Spatial Thought, Social Thought," 17–39.

165 **"taking a line for a walk"** Paul Klee, *Pedagogical Sketches* (New York:

Faber and Faber, 1968); E. S. Cross, A. F. Hamilton, and S. T. Grafton, "Building a Motor Simulation de Novo: Observation of Dance by Dancers," *NeuroImage* 31, no. 3 (2006): 1257–67.

168 **"A building is a struggle"** Kahn, quoted in H. F. S. Cooper, "The Architect Speaks," *Yale Daily News,* November 6, 1953, 2.

174 **in his Town Library in Viipuri** On Aalto's humanizing "rationalism" see my "Aalto's Embodied Rationalism," previously cited.

177 **such as Christopher Alexander** Such positions can be found in Christopher Alexander, *A Pattern Language*; Alexander, *The Nature of Order: An Essay on the Art of Building and the Nature of the Universe,* Books I–IV (Berkeley, CA: Center for Environmental Structure, 2002); Andreas Duany, Elizabeth Plater-Zyberk, and Jeff Speck, *Suburban Nation: The Rise of Sprawl and the Decline of the American Dream* (New York: North Point Press, 2000).

CHAPTER 5: PEOPLE EMBEDDED IN SOCIAL WORLDS

196 **notion of action settings** Roger Barker, *Ecological Psychology: Concepts and Methods for Studying the Environment of Human Behavior* (Stanford: Stanford University Press, 1968). On Barker, see also Ariel Sabar, *The Outsider: The Life and Times of Roger Barker* (Amazon, 2014); Phil Schoggen, *Behavior Settings: A Revision and Extension of Roger G. Barker's "Ecological Psychology"* (Stanford: Stanford University Press, 1989). Barker, coming out of the behaviorial psychology of the early postwar era, used the term "behavior setting." Because of behaviorism's deterministic connotations (or meanings!), I prefer the term "action setting," to emphasize the agency of humans, who make choices within the environmental settings they encounter.

197 **"variability in behavior"** Barker, *Ecological Psychology*, 4.

198 **So the Midwest Psychological Field Station** An excellent up-to-date account of how human evolution relied on the establishment of homes in settlements is by neuroanthropologist John S. Allen: *Home: How Habitat Made Us Human* (New York: Basic Books, 2015), especially 13–116. On the psychologically deranging effects of solitary confinement, see pbs.org/wgbh/pages/frontline/criminal-justice/locked-up-in-america/what-does-solitary-confinement-do-to-your-mind/; Mark Binelli, "Inside America's Toughest Federal Prison," *The New York Times Magazine,* March 29, 2015: 26–41, 56, 59.

199 **territory becomes a *place*** Maria Lewicka, "Place Attachment: How Far Have We Come in the Last 40 Years?" *Journal of Environmental Psychology* 31, no. 3 (2011): 218.

200 **And unless a person** U.S. Department of Housing and Urban Development, Office of Community Planning and Development, *The 2013 Annual Homeless Assessment Report (AHAR) to Congress.*

200 **place where you** Rebecca Solnit, *Storming the Gates of Paradise: Land-scapes for Politics* (Berkeley: University of California Press, 2008); 167.

201 **As we've seen** Lewicka, "Place Attachment," 207–30; Gifford, *Environmental Psychology*, 236–38; Irving Altman and Martin M. Chemers, *Culture and Environment* (Monterey, CA: Brooks/Cole, 1980); Judith Sixsmith, "The Meaning of Home: An Exploratory Study of Environmental Experience," *Journal of Environmental Psychology* 6, no. 4 (1986): 281–98; D. G. Hayward, "Home as an Environmental and Psychological Concept," *Landscape* (1975): 2–9; S. G. Smith, "The Essential Qualities of a Home," *Journal of Environmental Psychology* 14, no. 1 (1994): 31–46.

202 **As children** John Zeisel, *Inquiry by Design*, 356.

202 **The stories and narratives** Kaveli M. Korpela, "Place Attachment," *Oxford Handbook of Environmental and Conservation Psychology*, 148–63; Gifford, *Environmental Psychology*, 271–74; Zeisel, *Inquiry*, 147–150; Setha M. Low, "Cross-Cultural Place Attachment: A Preliminary Typology," in ed. Y. Yoshitake et al., *Current Issues in Environment-Behavior Research* (Tokyo: University of Tokyo, 1990).

202 **The four walls plus roof** Rhoda Kellogg, *Analyzing Children's Art* (New York: Mayfield, 1970); Kellogg collected 2,951 drawings of "home" by children from around the world.

203 **People's schemas of domestic** Sally Augustin, *Place Advantage*, 69–88.

203 **The strength of our affiliation** Lewicka, "Place Attachment," 218–24, writes, "We still know very little about the *processes* through which people become attached to places," and correctly notes that studies of place in environmental psychology have so overemphasized social processes that the physical attributes contributing to place attachment have been largely ignored: there is "a sad lack of theory that would connect people's emotional bonds with the physical side of places." By contrast, see Joanne Vining and Melinda S. Merrick, "Environmental Epiphanies: Theoretical Foundations and Practical Applications," *Oxford Handbook of Environmental and Conservation Psychology*, 485–508. The story of McDowell is told in Montgomery, *Happy City*, 106–45.

205 **"situationally normative"** Barker, *Ecological Psychology*, 34–35. The same is true for affordances: An atlas of the geography of the world can be a source of quiet entertainment and learning—but it can also be a decorative object on a coffee table, or a platform for writing or a cup of coffee, or a doorstop: Gibson, *Ecological Approach*, 37–38.

CHAPTER 6: DESIGNING FOR HUMANS

221 **Recognizing and identifying** Sternberg, *Healing Spaces*, 25–52; Chatterjee and Vartanian, "Neuroaesthetics," discuss differences between the human "liking" and "wanting" systems: liking is associated with opiates and cannabinoids, wanting with dopamine.

221 **Presumably, the functional** Thomas Albright, "Neuroscience for Architecture," in ed. Robinson and Palasmaa, *Mind in Architecture*, 197–217.

223 **The physics of materials** Sternberg, *Healing Spaces*, 25–52. On the other hand, the resonance of these mathematical figures with our visual system remains much in debate. See, for example, in Colin Ellard, *Places of the Heart*, a skeptical account of the putative human preference for fractal geometries (Ellard argues that our visual systems are far more attuned to contours in rapid gist identification). Similarly, on the prevalence of the golden ratio in nature and in architecture, and people's preferences for it, there remains vigorous debate. Adrian Bejan hypothesizes that people's gravitation to the golden ratio may be no more complex than that, at a large scale, a rectangle proportioned through the golden ratio fits most comfortably into the human cone of vision: "The Golden Ratio Predicted: Vision, Cognition, and Locomotion as a Single Design in Nature," *International Journal of Design and Nature and Ecodynamics* 4, no. 2 (2009): 97–104.

229 **"Good symmetry"** Kandel, *Insight*, 379; Ramachandran, *Tell-Tale Brain*, 200, 234–37; on our innate attraction to symmetry discussed in this and the following paragraphs, see, among others, Randy Thornhill and Steven Gangestad, "Facial Attractiveness," *Trends in Cognitive Science* 3, no. 2 (1999): 452–60; Karen Dobkins, "Visual Environments for Infants and Children," presentation at ANFA Conference 2012, Salk Institute, La Jolla, California.

233 **That is why generations** Jan Gehl, Lotte Johansen Kaefer, and Solvejg Reigstad, "Close Encounters with Buildings," *Urban Design International* 11 (2006): 29–47, quoted in Colin Ellard's wonderful chapter, "Boring Places," *Places of the Heart*, 107–24.

233 **This is also why** modernism was vilified in the public eye: not because there was anything inherently wrong with the various aesthetic languages its practitioners proposed, but because one version of it, techno-rationalism, was more widely adopted and very often the resulting designs were poorly conceived and very badly executed. For a more complex and forgiving view of modernism, see Sarah Williams Goldhagen, "Something to Talk About: Modernism, Discourse, Style," *Journal of the Society of Architectural Historians* 64, no. 2 (2005): 144–67.

236 **Another way to conceptualize** Chatterjee and Vartanian, "Neuroaesthetics," *Trends in Cognitive Science*; Ramachandran, *Tell-Tale Brain*, 231–33; Semir Zeki, *Inner Vision: An Exploration of Art and the Brain* (New York: Oxford University Press, 2000); Dzbic, Perdue, and Ellard, "Influence of Visual Perception on Responses in Real-World Environments," video (on YouTube), Academy of Neuroscience for Architecture conference, 2012.

242 **complexity to this patterned object through defamiliarization** On defamiliarization, see also Sarah Williams Goldhagen, *Louis Kahn's Situated Modernism* (New Haven: Yale University Press, 2001) 199–215.

246 **Change can be designed** These factors are somewhat (but not entirely) anal-

ogous to what Simon Unwin calls the "modifiers" of architecture in *Analyzing Architecture*, 3rd ed. (New York: Routledge, 2009), 43–56.

259 **They corral us** Hans-Georg Gadamer, in *Truth and Method* (New York: Continuum, 1975), writes that a person's engagement with a work of literature begins when we are "pulled up short" by the text because it "does not yield any meaning" or violates our expectations, 270. Years later, Semir Zeki discussed the neurological underpinnings of such engagement in his *Inner Vision: An Exploration of Art and the Brain*, writing that ambiguity in art activates our creative imagination, 25–28.

CHAPTER 7: FROM BLINDSIGHT TO INSIGHT

273 **Two quick examples** Bargh, "Embodiment in Social Psychology," 11; Augustin, *Place Advantage*, 10.

274 **a whole new literature** For an example, see Marc Augé, *Non-Places: An Introduction to Supermodernity* (New York: Verso, 2009).

275 **One large part** In 1994, the Carnegie Task Force reported that children who grow up in experientially impoverished environments reliably suffer permanent cognitive setbacks in comparison with those raised in enriched environments: quoted in Michael Mehaffy and Nikos Salingaros, "Science for Designers: Intelligence and the Information Environment," *Metropolis*, February 25, 2012: metropolismag.com/Point-of-View/February-2012/Science-for-Designers-Intelligence-and-the-Information-Environment/. Mehaffy and Salingaros's series in *Metropolis* covers many issues of interest, including fractals and biophilia.

276 **A good starting point** Martha Nussbaum, *Creating Capabilities: The Human Development Approach* (Cambridge, MA: Harvard University Press, 2011).

281 **"having decent, ample housing"** Nussbaum, *Creating Capabilities*, loc. 466 Kindle edition.

281 **Built environments that accord** Gerd Kempermann, H. Georg Kuhn, and Fred Gage, "More Hippocampal Neurons in Adult Mice Living in an Enriched Environment," *Nature* 386, no. 6624 (April 1997): 493–95; Alessandro Sale et al., "Enriched Environment and Acceleration of Visual System Development," *Neuropharmacology* 47, no. 5 (2004): 649–60; Matthew Dooley and Brian Dooley, ANFA lecture, anfarch.org/activities/Conference2012Videos.shtml; Rusty Gage, ANFA lecture, anfarch.org/activities/Conference2012Videos.shtml; Kevin Barton, ANFA lecture, anfarch.org/activities/Conference2012Videos.shtml. On cognitive deficits resulting from early childhood development in deprived environments, see James Heckman, Rodrigo Pinto, and Peter Savelyev, "Understanding the Mechanisms Through Which an Influential Early Childhood Program Boosted Adult Outcomes," *American Economic Review* 103, 6 (2013): 2052–86.

285 **but as far as we know** Antonio Damasio, *Self Comes to Mind*, 67–94.

290 **We cannot but move around** Linda B. Smith, "Action Alters Shape Categories," *Cognitive Science 29* (2005): 665–79; Linda B. Smith, "Cognition as a Dynamic System: Principles from Embodiment," *Developmental Review* 25 (2005): 278–98; Linda B. Smith and Esther Thelen, "Development as a Dynamic System," *Trends in Cognitive Science* 7, no. 8 (2003): 343–48: All these articles demonstrate that shape perception is a dynamic process that requires actual manipulation as well as simulated movement.

Index

Page numbers in italics refer to illustrations.

About the Author

SARAH WILLIAMS GOLDHAGEN taught at Harvard University's Graduate School of Design for ten years and was the *New Republic*'s architecture critic until recently. Currently a contributing editor at *Art in America* and *Architectural Record*, she is an award-winning writer who has written for many national and international publications, including the *New York Times*, the *American Prospect*, and *Harvard Design Magazine*. She lives in New York City.